PENGUIN BOOKS

WE ARE THE WEATHER MAKERS

Tim Flannery is an internationally acclaimed scientist, explorer and writer. As a field zoologist he has discovered and named more than thirty new species of mammals (including two tree-kangaroos) and at thirty-four he was awarded the Edgeworth David Medal for Outstanding Research.

Formerly director of the South Australian Museum, Tim is chairman of the South Australian Premier's Science Council and Sustainability Roundtable; a director of the Australian Wildlife Conservancy; a leading member of the Wentworth Group of Concerned Scientists; and the National Geographic Society's representative in Australasia. In April 2005 he was honoured as Australian Humanist of the Year. He will take up a position at Sydney's Macquarie University in 2007.

Tim's books include the award-winning international bestsellers *Country*, *The Eternal Frontier* and *Throwim Way Leg*. He has also edited and introduced many historical ... ney, *The Diaries of*

WE ARE THE WEATHER MAKERS

THE STORY OF GLOBAL WARMING

TIM FLANNERY

PENGUIN BOOKS

Go to **www.theweathermakers.com**
for Notes for Teachers and Students

PENGUIN BOOKS

Published by the Penguin Group
Penguin Books Ltd, 80 Strand, London WC2R ORL, England
Penguin Group (USA) Inc., 375 Hudson Street, New York, New York 10014, USA
Penguin Group (Canada), 90 Eglinton Avenue East, Suite 700, Toronto, Ontario, Canada M4P 2Y3 (a division of Pearson Penguin Canada Inc.)
Penguin Ireland, 25 St Stephen's Green, Dublin 2, Ireland (a division of Penguin Books Ltd)
Penguin Group (Australia), 250 Camberwell Road, Camberwell, Victoria 3124, Australia (a division of Pearson Australia Group Pty Ltd)
Penguin Books India Pvt Ltd, 11 Community Centre, Panchsheel Park, New Delhi – 110 017, India
Penguin Group (NZ), 67 Apollo Drive, Rosedale, North Shore 0632, New Zealand (a division of Pearson New Zealand Ltd)
Penguin Books (South Africa) (Pty) Ltd, 24 Sturdee Avenue, Rosebank, Johannesburg 2196, South Africa

Penguin Books Ltd, Registered Offices: 80 Strand, London WC2R ORL, England

www.penguin.com

First published in Australia by Text Publishing Company 2006
First published in Great Britain in Penguin Books 2007
1

Printed in England by Clays Ltd, St Ives plc

978-0-141-03407-2

www.greenpenguin.co.uk

Penguin Books is committed to a sustainable future for our business, our readers and our planet. The book in your hands is made from paper certified by the Forest Stewardship Council.

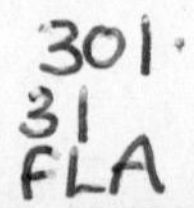

To David and Emma, Tim and Nick, Noriko and Naomi, Puffin and Galen, Will, Alice, Julia and Anna, and of course Kris, with love and hope; and to all of their generation who will have to live with the consequences of our decisions.

CONTENTS

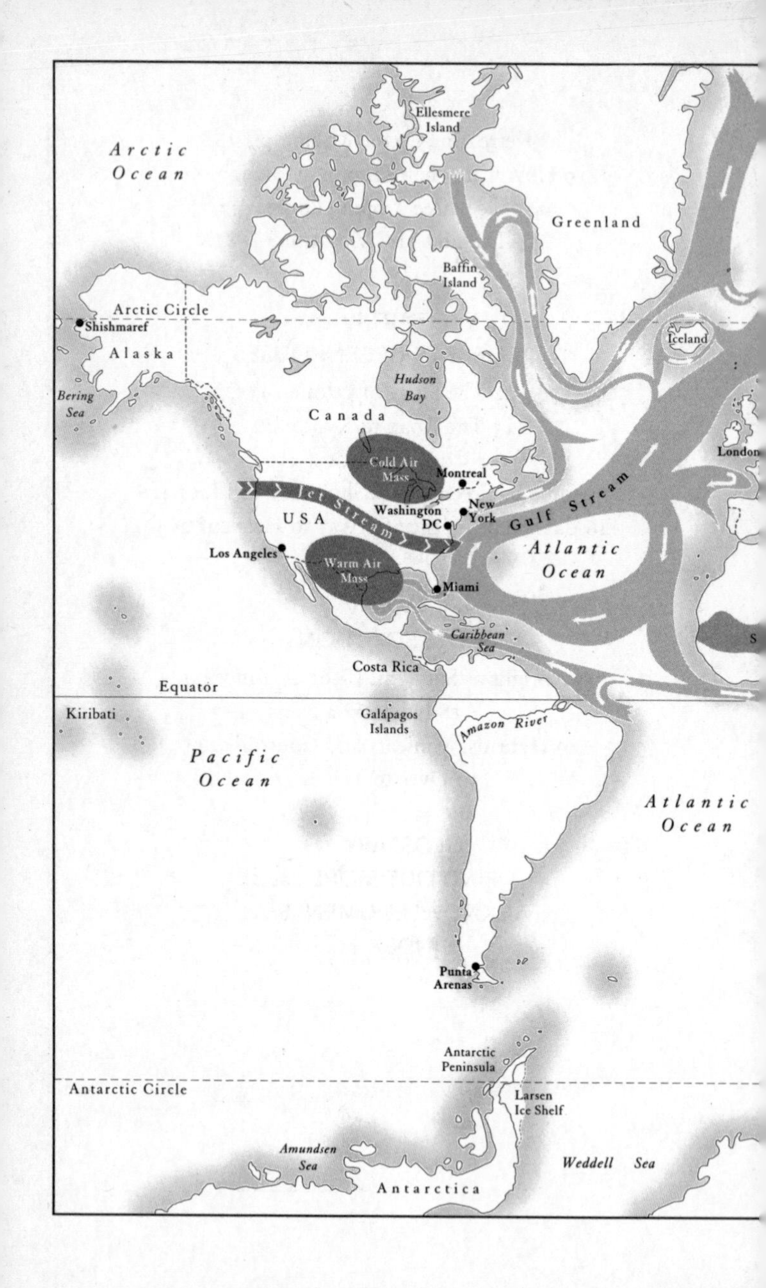
Arctic Ocean
Ellesmere Island
Greenland
Baffin Island
Arctic Circle
Shishmaref
Alaska
Iceland
Hudson Bay
Bering Sea
Canada
London
Cold Air Mass
Montreal
Jet Stream
New York
Washington DC
USA
Gulf Stream
Los Angeles
Warm Air Mass
Atlantic Ocean
Miami
Caribbean Sea
Costa Rica
Equator
Kiribati
Galápagos Islands
Amazon River
Pacific Ocean
Atlantic Ocean
Punta Arenas
Antarctic Peninsula
Antarctic Circle
Larsen Ice Shelf
Amundsen Sea
Weddell Sea
Antarctica

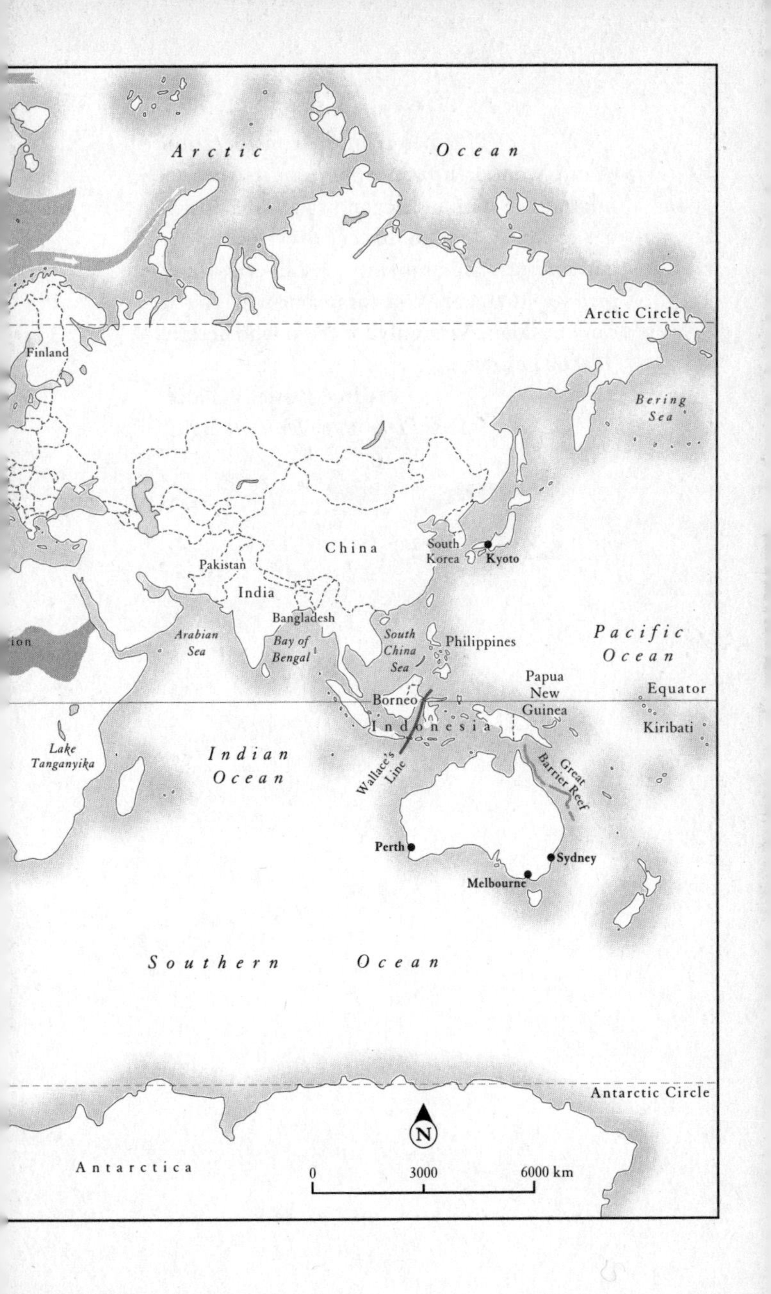

Arctic
Ocean
Arctic Circle
Finland
Bering
Sea
China
South
Korea
Kyoto
Pakistan
India
Bangladesh
Arabian
Sea
Bay of
Bengal
South
China
Sea
Philippines
Pacific
Ocean
Papua
New
Guinea
Equator
Borneo
Indonesia
Kiribati
Lake
Tanganyika
Indian
Ocean
Wallace's
Line
Great
Barrier Reef
Perth
Sydney
Melbourne
Southern
Ocean
Antarctic Circle
N
Antarctica
0
3000
6000 km

It has been the consideration of our wonderful atmosphere in its various relations to human life, and to all life, which has compelled me to this cry for the children and for outraged humanity . . . Let everything give way to this . . . Vote for no one who says 'it can't be done'. Vote only for those who declare 'It shall be done'.

Alfred Russel Wallace,
Man's Place in the Universe, 1903.

INTRODUCTION: WHAT IS CLIMATE CHANGE?

Anyone picking up this book might wonder about its title. *We Are the Weather Makers* is a serious thing to say. And if anyone had said to me a decade ago that our planet was in urgent danger I wouldn't have paid much attention. The story of this book is what I've learned since then, and how I came to change my view.

In the last decade climate science has undergone a revolution, and now we understand a great deal more about Earth's climate system and how it is changing. Climate is always changing, of course, but it is now doing so at an unnatural pace, and we are causing it. Unfortunately, most of those changes will damage our world.

I've written this book in the hope that people can continue to have the chance, as I did, to stand on a glacier high atop a tropical mountain, and to look way down over dense jungles, plains and mangrove swamps, and finally see tropical reefs in the distance.

It should be everyone's birthright to experience our wonderful planet to the fullest, to have the chance to see polar bears, great whales and Antarctic glaciers in real life.

I believe that it's profoundly wrong to deprive future generations of this just so we can continue wasting electricity and driving oversize cars.

And I want to empower readers: our leaders in politics and business need to hear your voice. I hope this book helps you to act firmly, because if you let them continue doing things in the same old ways you will become part of their failure.

In 1981, when I was in my mid-twenties, I climbed Mt Albert Edward, one of the highest peaks on the tropical island of New Guinea.

Tree-fern grassland in the Star Mountains, central New Guinea. It was this habitat I noticed disappearing under encroaching forests, as a result of global warming.

The bronzed grasslands of the summit were a contrast to the green jungle all around, and among the alpine tussocks grew groves of tree-ferns, whose lacy fronds waved above my head.

Downslope, the tussock grassland ended abruptly at a stunted, mossy forest. A single step could carry you from sunshine into the gloom, where the pencil-thin saplings were covered with moss, lichens and filmy ferns.

In the leaf litter on the forest floor I was surprised to find the trunks of dead tree-ferns. Tree-ferns grew only in the grassland, so it was clear that the forest was climbing towards the mountain peak. I guessed it had swallowed at least thirty metres of grassland in less time than it takes for a tree-fern to rot on the damp forest floor—a decade or two at most.

Why was the forest expanding? I remembered reading that New Guinea's glaciers were melting. Had the temperature on Mt Albert Edward warmed enough to permit trees to grow where previously only grasses could take root? Was this evidence of climate change?

I am a palaeontologist, someone who studies fossils and geological periods, so I know how important changes in climate have been in determining the fate of species. But this was the first evidence I'd seen that it might affect Earth during my lifetime. I knew there was something wrong, but not quite what it was.

Despite the good position I was in to understand the

significance of these observations, I soon forgot about them. What seemed like more urgent issues demanded my attention. Rainforests were being felled for timber and to make agricultural land, and the larger animal species living there were being hunted to extinction. In my own country of Australia, rising salt was threatening to destroy the most fertile soils. Overgrazing, water pollution and the logging of forests all threatened precious ecosystems and biodiversity—the range and variety of life forms that exist in our environment.

So is climate change a huge threat, or nothing to worry about? Or is it something in between—an issue that we must soon face, but not yet?

Even scientists don't agree on every aspect of climate change research. We are trained sceptics, always questioning our own and others' work. A scientific theory is only valid for as long as it has not been disproved. And climate change can be difficult for many people to think about calmly because it arises from so many things we take for granted in the way we live.

Some things about climate change are certain. It results from a special kind of air pollution. We know exactly the size of our atmosphere and the volume of pollutants pouring into it. The story I want to tell here is about the impacts of some of those pollutants (known as greenhouse gases) on all life on Earth.

For the last 10,000 years Earth's thermostat, or climate

control, has been set to an average surface temperature of around 14°C. On the whole this has suited human beings splendidly, and we have been able to organise ourselves in a most impressive manner—planting crops, domesticating animals and building cities.

Finally, over the past century, we have created a truly global civilisation. Given that in all of Earth's history the only other creatures able to organise themselves on a similar scale are ants, bees and termites—which are tiny in comparison to us and have small resource requirements—this is quite an achievement.

Earth's thermostat is a complex and delicate mechanism, at the heart of which lies carbon dioxide (CO_2), a colourless and odourless gas formed from one carbon and two oxygen atoms.

CO_2 plays a critical role in maintaining the balance necessary to all life. It is also a waste product of the fossil fuels—coal, oil and gas—that almost every person on the planet uses for heat, transport or other energy needs. On dead planets such as Venus and Mars, CO_2 makes up most of the atmosphere, and it would do so here if living things and Earth's processes didn't keep it within bounds. Our planet's rocks, soils and waters are packed with carbon atoms itching to combine with oxygen and get airborne. Carbon is everywhere.

For the past 10,000 years CO_2 has made up around three parts per 10,000 in Earth's atmosphere. That's a small amount—0.03 per cent—yet it has a big influence on temperature. We create CO_2 every time we burn fossil fuels to drive a car, cook a meal or turn on a light, and the gas we create lasts around a century in the atmosphere. So the proportion of CO_2 in the air we breathe is now rapidly increasing, and this is causing our planet to warm.

The matanim cuscus. This woman's husband caught the rare creature in the forests of central New Guinea in 1985. It may well already be extinct as a result of climate change.

By late 2004, I was really worried. The world's leading science journals were full of reports that glaciers were melting ten times faster than previously thought, that atmospheric greenhouse gases had reached levels not seen for millions of years, and that species were vanishing as a result of climate change. There were also reports of extreme weather events, long-term droughts and rising sea levels.

We cannot wait for someone to solve this problem of carbon emissions for us. We can all make a difference and help combat climate change at almost no cost to our lifestyle. And in this, climate change is very different from other environmental issues such as biodiversity loss or the ozone hole.

The best scientific evidence indicates that we need to reduce our CO_2 emissions by 70 per cent by 2050.

How can we do this?

If your family owns a four-wheel-drive and replaces it with a hybrid fuel car, which combines an electric motor with a petrol-driven engine, you can instantly cut your transport emissions by 70 per cent.

If your home's electricity provider offers a green option, you will be able to make equally major cuts in your household emissions for the daily cost of an ice cream. Just ask for your power to come from renewable energy sources such as wind, solar or hydro.

And if you encourage your family and friends to vote for a politician who has a deep commitment to reducing CO_2 emissions, you might change the world.

We have all the technology we need to change to a carbon-free economy. All we need is to apply our knowledge and develop our understanding. The main thing stopping us going forward are the pessimism and confusion created by people who want to go on polluting so that they can make money.

Our future depends on readers like you. Whenever my family gathers for a special event, the true scale of climate change is never far from my mind. My mother, who was born when motor vehicles and electric lights were still novelties, glows in the company of her grandchildren, some of whom are not yet ten.

To see them together is to see a chain of the deepest love that spans 150 years, for those grandchildren will not reach my mother's present age until late this century. To me, to her, and to their parents, their welfare is every bit as important as our own.

Climate change affects almost every family on this planet. Seventy per cent of all people alive today will still be alive in 2050.

I

THE ATMOSPHERE

ONE

EVERYTHING IS CONNECTED

Until a black mood takes her and she rages about our heads, most of us are unaware of our atmosphere.

The 'atmosphere': what a dull word for such a wondrous thing. In 1903 Alfred Russel Wallace, co-founder with Charles Darwin of the theory of evolution by natural selection, came up with the phrase 'The Great Aerial Ocean' to describe the atmosphere. It's a far better name, because it conjures in the mind's eye the currents and layers that create the weather far above our heads, and which is all that stands between us and the vastness of space.

Wallace lived in a romantic era of science. Back then discoveries about the atmosphere were as exciting as dredging up monsters from the deep or seeing pictures sent from Mars. What an astonishing thing it is, Wallace mused, that without dust sunsets would be as dull as dishwater and shadows would be as impenetrable to our eyes as concrete.

The atmosphere is amazing. It protects all life, connects everything with everything else, and has regulated our planet's temperature for nearly 4 billion years.

Over time Earth has become better at regulating its temperature. For nearly half of its existence—from 4 billion to around 2.2 billion years ago—Earth's atmosphere would have been deadly to creatures such as us. Back then all life was microscopic—algae and bacteria—and its hold on our planet was weak.

By around 600 million years ago oxygen levels had increased enough to permit the survival of larger creatures—ones whose fossils can be seen with the unaided eye. These early organisms lived during a period of momentous climate change, when four major ice ages gripped the planet. Six hundred million years ago, for example, Earth froze all the way to the equator. Just a few living things sheltered in refuges under the equatorial ice.

The deep freeze was aided by a powerful mechanism known as Earth's albedo. *Albedo* is Latin for 'whiteness', and of course a snow-covered Earth is a lot whiter than one that isn't. Why does this matter? One-third of all energy reaching Earth from the Sun is reflected back to space by white surfaces. Fresh snow reflects 80–90 per cent of light while water reflects only 5–10 per cent.

Once a certain proportion of the planet's surface is bright ice and snow, a runaway cooling effect is created which freezes it completely.

That threshold is crossed when ice sheets start heading towards the equator and reach around 30 degrees of latitude, as far south as Shanghai or New Orleans.

The big freeze of 600 million years ago lasted for millions of years. But around 540 million years ago living things began to build skeletons of carbonate. They did this by absorbing CO_2 from sea water. This increased carbon dioxide levels in the atmosphere and, ever since, ice ages have been rare. There have only been two: between 355 and 280 million years ago; and now, for the past 33 million years.

Other changes were occurring that would also have a profound impact on Earth's thermostat. This was during the Carboniferous Period, when forests first covered the land, and when most of the coal deposits that now feed our industry were laid down. All of the carbon in that coal was once tied up in CO_2 floating in the atmosphere, so those primitive forests must have had an enormous influence on the carbon cycle.

Other creatures have influenced the carbon cycle more recently. The spread of modern coral reefs after around 55 million years ago drew unimaginable volumes of CO_2 from the atmosphere, further altering the climate, perhaps cooling it.

The evolution and spread of grasses around 6–8 million years ago may have changed things again. Grasses contain much less carbon than do forests. They also absorb less sunlight (having a different albedo), and produce less water vapour, which affects cloud formation.

Another likely influence is the elephant, a great destroyer

of forests. Like humans, its original homeland was Africa, and as it spread across the planet around 20 million years ago (only Australia escaped colonisation) it too must have affected the carbon cycle. Exactly what happened to the climate as a result of these changes is not known, but it seems certain that the activities of these animals and plants altered the atmosphere in subtle ways.

When it comes to climate, everything is connected to everything else. To understand what might happen in the future we need to know as much as possible about our atmosphere and how it has operated in the past.

TWO

THE GREAT AERIAL OCEAN

We have all heard the terms greenhouse gases, global warming and climate change.

Greenhouse gases trap heat near Earth's surface. As they increase in the atmosphere, the extra heat they trap leads to global warming. This warming in turn places pressure on Earth's climate system, and can lead to climate change.

There is a difference between weather and climate. Weather is what we experience each day. Climate is the sum of all weathers over a certain period, for a region or for the planet as a whole.

The atmosphere has four distinct layers, which are defined on the basis of their temperature and the direction of their temperature gradient.

The lowest part of the atmosphere is known as the troposphere. Its name means the region where air turns over, and it is so called because of the vertical mixing of air that occurs there.

The troposphere extends on average to twelve kilometres above the Earth's surface, and it contains 80 per cent

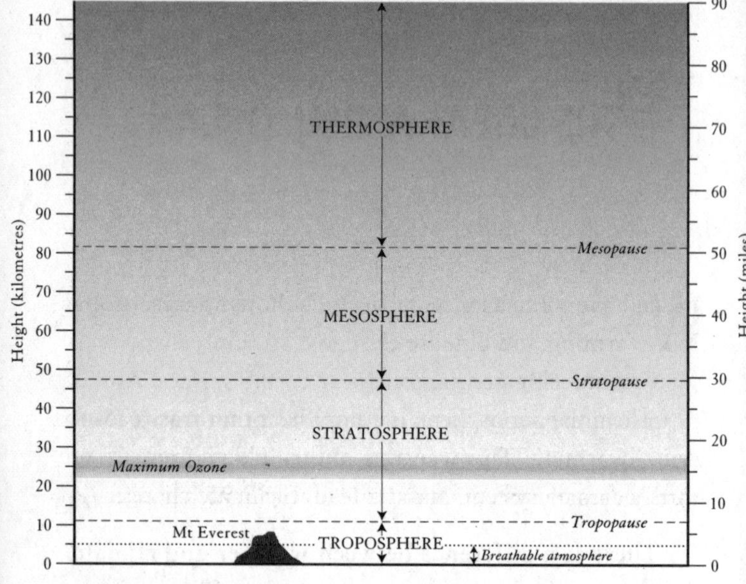

The four major parts of the atmosphere, and their associated boundaries. Only a small part of the troposphere is breathable air.

of all the atmosphere's gases. Its bottom third is the only part of the entire atmosphere that is breathable.

The key thing about the troposphere is that it is 'upside down'—it is warmest at the bottom, and cools by 6.5°C per vertical kilometre travelled. And it is the only portion of the atmosphere whose northern and southern halves (divided by the equator) hardly mix. That's why the inhabitants of the Southern Hemisphere don't breathe the polluted air that limits horizons and dulls panoramas in the more populated north.

The next layer of the atmosphere, known as the stratosphere, meets the troposphere at the tropopause. The stratosphere gets hotter as one rises through it. It is distinctly layered, and fierce winds circulate within it.

Some fifty kilometres above the surface of Earth lies the mesosphere. At –90°C it's the coldest portion, and above it lies the final layer, the thermosphere, which is a thin dribble of gas extending far into space. There temperatures can reach 1000°C, yet because the gas is so thinly dispersed it would not feel hot to the touch.

The great aerial ocean is composed of nitrogen (78 per cent), oxygen (20.9 per cent) and argon (0.9 per cent). These three gases comprise almost all—over 99.95 per cent—of the air we breathe.

The atmosphere's capacity to hold water (H_2O) depends on its temperature: at 25°C water vapour makes up 3 per cent of what we inhale. But it's the minor elements, which scientists refer to as 'trace' gases—the remaining one-twentieth of 1 per cent—that spice the mix, and some of them are vital to life on this planet.

Take, for example, ozone. Its molecules are composed of three oxygen atoms. Ozone makes up just ten molecules of every million tossed about in the currents of the great aerial ocean. Yet without the shielding effect of that 0.001 per cent, we would soon go blind, die of cancer or succumb to any number of other problems caused by ultra-violet radiation.

We are so small, and the great aerial ocean so vast, it seems impossible that we could do anything to affect it. But if we were to imagine Earth as an onion, our atmosphere would be no thicker than its outermost parchment skin. Its breathable portion does not even completely cover the surface of the planet—which is why climbers on Mt Everest must wear oxygen masks.

The atmosphere looks big because it is made of gas, but if we could compress that gas to liquid we would discover that the atmosphere is only one five-hundredth the size of the oceans. That's why humanity's major environmental problems—the ozone hole, acid rain, and climate change—result from air pollution.

And yet the atmosphere is dynamic. The air you just exhaled has already spread far and wide. The CO_2 from a breath last week may now be feeding a plant on a distant continent, or plankton in a frozen sea.

In a matter of months all of the CO_2 you just exhaled will have dispersed around the planet.

The atmosphere is also telekinetic, which means changes can occur in it simultaneously in different regions. It can transform itself from one climatic state into another instantly. This allows storms, droughts, floods or wind patterns to alter on a global level, at more or less the same time.

Because communication around the planet is now instantaneous, our global civilisation is telekinetic as well, which is why it is such a powerful force. But its telekinesis also explains why regional disruptions—such as wars, famines and diseases—can have dire consequences for humanity as a whole.

The atmosphere blocks out most forms of radiative energy. Many of us imagine that daylight is the only energy we receive from the Sun, but sunlight—visible light—is only a small part in a broad spectrum of radiation that the Sun shoots our way.

The greenhouse gases especially block the forms of radiative energy we call heat. By doing this, however, these gases become unstable and eventually release the heat, some of which radiates back to Earth. Greenhouse gases may be rare, but their impact is massive. They warm our world and, by trapping more heat near the surface, account for the 'upside down' troposphere.

Some idea of the power greenhouse gases have to influence temperature can be gained from other planets. The atmosphere of Venus is 98 per cent CO_2, and its surface temperature is 477°C.

If CO_2 made up 1 per cent of the atmosphere, it could bring the surface temperature of our planet to boiling point.

If you want to feel how greenhouse gases work, visit New York in August. The unhealthy heat and humidity leave you in a lather of sweat in a crowded environment of concrete, parched bitumen and sticky human bodies. And the worst of it comes at night, when humidity and a thick layer of cloud lock in the heat. I recall tossing and turning between sweat-soaked sheets in a room in a rundown neighbourhood known for its drug-dealers and addicts. As my eyes became gritty and my skin began to crust up, I could smell the grime of the city's 8 million human bodies.

I longed to be in a desert—a dry, clear desert where, no matter how hot the day, the clear skies of night bring blessed relief. The difference between a desert and New York City at night is a single greenhouse gas—the most powerful of them all—water vapour. Remembering that water vapour retains two-thirds of all the heat trapped by all the greenhouse gases, I cursed the clouds overhead.

But clouds have a saving grace too. Unlike the other greenhouse gases, water vapour in the form of clouds blocks part of the Sun's radiation by day, keeping temperatures down.

So how do CO_2 and water vapour interact? As the concentration of CO_2 increases, it warms the atmosphere just a little, which allows it to retain more water vapour. This in turn magnifies the original warming. You can think of CO_2 as the lever that shifts our climate—or the match that lights the climate change firestorm.

CO_2 is produced whenever we burn something or when things decompose. So how do we measure it? In the 1950s, a climatologist named Charles Keeling climbed Mt Mauna

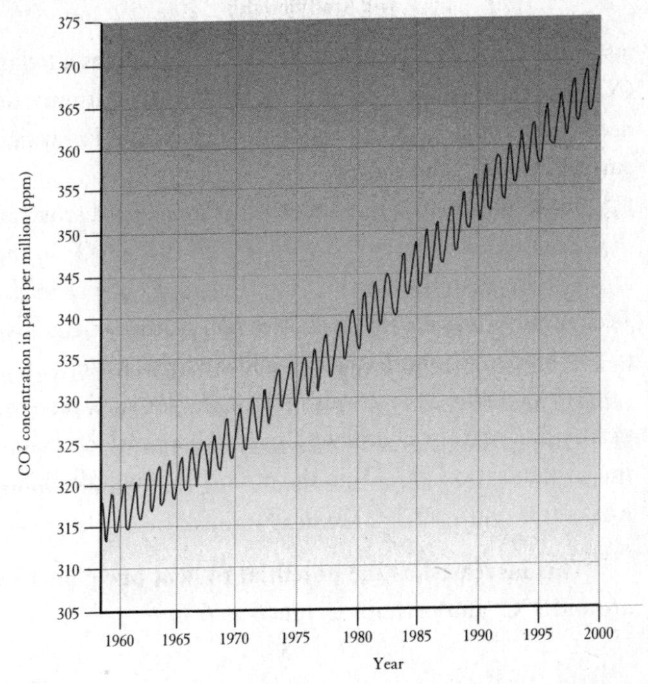

The Keeling curve shows the concentration of CO_2 in the atmosphere as measured atop Mt Mauna Loa, Hawaii, between 1958 and 2000. The saw-tooth effect results from the seasonal changes in northern forests, but the inexorable rise is due to the burning of fossil fuels.

Loa in Hawaii to record CO_2 concentrations in the atmosphere. From this he created a graph, known as the Keeling curve, that is one of the most wonderful things I've ever seen. In it you can see our planet breathing.

Every northern spring as the sprouting greenery extracts CO_2 from the atmosphere our Earth begins a great inspiration, which is recorded on Keeling's graph as a fall in CO_2 concentration. Then, in the northern autumn, as decomposition generates CO_2, there is an exhalation which enriches the air with the gas.

But Keeling's work revealed another trend. He discovered that each exhalation ended with a little more CO_2 in the atmosphere than the one before. This innocent rise in the Keeling curve was the first definitive sign that we might have to pay a price for our fossil-fuel-addicted civilisation.

Try extending the graph's trajectory forward into the twenty-first century. Unless we change the way we do things, the concentration of CO_2 in the atmosphere will double—from three parts per 10,000 to six.

This increase has the potential to heat our planet by around 3°C, and perhaps as much as 6°C.

THREE

THE GREENHOUSE GASES

When scientists first realised that increasing CO_2 in the atmosphere was linked to climate change, some were puzzled. There was so little of it in the atmosphere—how could it change the climate of an entire planet? Then they discovered that CO_2 acts as a trigger for that potent greenhouse gas, water vapour.

Carbon dioxide also lasts a long time in the atmosphere: around 56 per cent of all the CO_2 that humans have liberated by burning fossil fuel in the past century is still aloft, which is the cause—directly and indirectly—of around 80 per cent of all global warming.

The fact that a known proportion of CO_2 remains in the atmosphere allows us to calculate, in round numbers, a carbon budget for humanity. We can do this in gigatonnes—a gigatonne is a billion tonnes. The carbon budget tells us how much more carbon we can put into the atmosphere before we trigger dangerous changes, which are widely acknowledged to occur at between 450–550 parts per million of CO_2.

Prior to 1800, the start of the Industrial Revolution,

there were about 280 parts per million of CO_2 in the atmosphere, which equates to around 586 gigatonnes of carbon. (To make comparisons easy, figures like this relate only to the carbon in the CO_2 molecule. The actual weight of the CO_2 would be 3.7 times greater.)

Today the figures are 380 parts per million or around 790 gigatonnes in total.

If we wish to stabilise CO_2 emissions below that threshold of dangerous change, we will have to limit all future human emissions to around 600 gigatonnes. Just over half of this will stay in the atmosphere, raising levels to around 1100 gigatonnes, or 550 parts per million, by 2100.

This will be a tough budget for humanity to abide by. Over a century, it equates to around 6 gigatonnes per year. Compare this with the average of 13.3 gigatonnes that accumulated each year throughout the 1990s (half of this from burning fossil fuel). And remember that the human population is set to rise from 6 billion now to 9 billion in 2050. You can see the problem.

Even in the long view, this rise in CO_2 is exceptional. Its concentration in the atmosphere in times past can be measured from bubbles of air preserved in ice. By drilling more than three kilometres into the Antarctic ice cap, scientists have drawn out an ice-core that spans almost a million years of Earth history.

The power of the ice core record to tell us about the climate and atmosphere in times past came home to me when I visited the University of Copenhagen's ice core store in Denmark. I had arrived straight from an Australian summer and the store was –26°C. The hardy Dane showing me about seemed unaware of my shock. Concern for my freezing nose instantly vanished, however, when my guide held out a cylindrical piece of ice around one metre long and pointed to a layer of ice in it around five centimetres thick. That ice, he said, fell as snow over central Greenland in the year Jesus was born, and the tiny specks I could see in it were bubbles of air trapped in the ice. From those bubbles scientists could tell what the levels of CO_2 and other atmospheric gases were that year, which reveals quite a lot about the state of the climate. The atmosphere mixes so quickly, he said, that it's possible those bubbles contain a few molecules breathed out by the Holy Family during that first year.

The unique record of the ice cores demonstrates that during cold times CO_2 levels have dropped to around 160 parts per million, and that until recently they never exceeded 280 parts per million. The Industrial Revolution, with its steam engines and smoky factories, changed that. By 1958, when Keeling began his measurements of CO_2 atop Mauna Loa, it comprised 315 parts per million.

It is our servants—the billions of engines that we have built to run on fossil fuels such as coal, petrol and oil-based fuels, and gas—that play the leading role in manufacturing CO_2. Most dangerous of all are the power plants that use coal to generate electricity. Black coal (anthracite) is composed of at least 92 per cent carbon, while dry brown coal is around 70 per cent carbon and 5 per cent hydrogen.

Some power plants burn through 500 tonnes of coal per hour. They are so inefficient that around two-thirds of the energy created is wasted. And to what purpose? Simply to boil water, which generates steam that moves the colossal turbines to create the electricity that powers our homes and factories.

Most of us have no idea that nineteenth-century technology makes our twenty-first-century gadgets work.

There are around thirty other greenhouse gases. Think of them as glass windows in a ceiling, each gas representing a different window. As the number of windows increases, more light is admitted into the room, where it is trapped as heat.

After CO_2, methane is the next most important greenhouse gas. Methane is created by microbes that thrive in oxygenless environments such as stagnant pools and bowels, which is why it abounds in swamps, farts and belches. It comprises just 1.5 parts per million of the atmosphere, but its concentration has doubled over the last few hundred years.

Methane is sixty times more potent at capturing heat energy than CO_2, but thankfully lasts fewer years in the atmosphere. It is estimated that methane will cause 15 to 17 per cent of all global warming experienced this century.

Nitrous oxide (laughing gas) is 270 times more efficient at trapping heat than CO_2. It is far rarer than methane but it lasts 150 years in the atmosphere. Around a third of our global emissions of this gas come from burning fossil fuels. The rest comes from burning biomass (plant and animal material) and the use of nitrogen-containing fertilisers. While there are natural sources of nitrous oxide, human emissions now greatly exceed them in volume. Today there is 20 per cent more nitrous oxide in the atmosphere than there was at the beginning of the Industrial Revolution.

The rarest of all greenhouse gases are members of the hydrofluorocarbon (HFC) and chlorofluorocarbon (CFC) families of chemicals. These products of human ingenuity did not exist before industrial chemists began to manufacture them. Some, such as the tongue-twisting dichlorotrifluoroethane, which was once used in refrigeration, are ten thousand times more potent at capturing heat energy than CO_2, and they can last in the atmosphere for many centuries. We shall meet this class of chemicals again later, when we come to the story of the ozone hole.

For the moment, though, we need to know more about the carbon in CO_2. Both diamonds and soot are pure forms of carbon; the only difference is how the atoms are arranged. Carbon is everywhere on the surface of planet Earth. It is constantly shifting in and out of our bodies as well as from rocks to sea or soils, and from there to the atmosphere and back again.

Were it not for plants and algae, we would soon suffocate in CO_2 and run out of oxygen. Through photosynthesis (the process whereby plants create sugars using sunlight and water) plants take our waste CO_2 and use it to make their own energy, creating a waste stream of oxygen along the way. It's a neat and self-sustaining cycle that forms the basis of life on Earth.

The volume of carbon circulating around our planet is enormous. Around a trillion tonnes of it is tied up in living things, while the amount buried underground is far, far greater. And for every molecule of CO_2 in the atmosphere, there are fifty in the oceans.

The places where the carbon goes when it leaves the atmosphere are known as carbon sinks. You and I and all living things are carbon sinks, as are the oceans and the soil and some of the rocks under our feet.

Over eons, much CO_2 has been stored in the Earth's crust. This occurs as dead plants are buried and carried underground, where they become fossil fuels. On a shorter time scale, a lot of carbon can be stored in soils, where it forms the black mould that gardeners love.

Even the belching of volcanoes (which contains much CO_2) can disturb the climate. And meteorites that collide with Earth can also disrupt the carbon cycle by upsetting the oceans, atmosphere and the crust.

Scientists know where the CO_2 that we produce goes.

This is because the gas derived from fossil fuels has a distinct chemical signature and can be tracked as it circulates around the planet. In very round figures, 2 gigatonnes is absorbed by the oceans and a further 1.5 gigatonnes is absorbed by life on land each year.

The contribution made by the land results partly from an accident of history—America's frontier phase of development. Mature plants, trees and forests don't take in much CO_2 for they are in balance, releasing CO_2 as old vegetation rots, then absorbing it as the new grows. The world's largest forests—the coniferous forests of Siberia and Canada—and the tropical rainforests don't absorb as much carbon as new forests.

During the nineteenth and early twentieth centuries, America's pioneers cut and burned the great eastern forests, and burned and grazed the western plains and deserts. Then shifts in land use allowed the vegetation to grow back. As a result, most of America's forests are fewer than sixty years old and are regrowing vigorously, absorbing around half a billion tonnes of CO_2 annually from the atmosphere. And remember, trees are built of air, not the ground they sprout from: timber, leaves and bark were once, not that long ago, CO_2 in the atmosphere.

Newly planted forests in China and Europe may be absorbing an equal amount. For a few crucial decades these young forests have helped cool our planet by absorbing excess CO_2.

But as the Northern Hemisphere's forests and shrublands recover from their damage at the hands of the pioneers, they will extract less and less CO_2, just when humans are pumping more of it into the atmosphere.

If we take a long term view, there really is only one major carbon sink on our planet—the oceans. They have absorbed 48 per cent of all the carbon emitted by humans between 1800 and 1994.

The world's oceans vary in their ability to absorb carbon. One ocean basin alone, the North Atlantic—which comprises 15 per cent of the ocean surface—contains almost a quarter of all the carbon emitted by humans since 1800. Shallow seas are a carbon kidney and have removed 20 per cent of all carbon dioxide emitted by humans.

Scientists are worried that changes in ocean circulation brought about by global warming might degrade the effectiveness of this 'carbon kidney'. There are many ways that this could happen, one of which you can see in a warm can of soft drink. That fizz on opening the can fades—indicating that the liquid has quickly released the CO_2 that gives it its sparkle. Cold drinks hold their fizz longer. Cold sea water can hold more carbon than warm sea water, so as the ocean warms it becomes less able to absorb the gas.

Sea water also contains carbonate. It reaches the oceans from rivers that have flowed over limestone or other lime-containing rocks, and it reacts with the CO_2 absorbed into the oceans. At present there is a balance between carbonate concentration and the CO_2 absorbed. As the CO_2 concentration increases in the oceans, however, the carbonate is being used up.

The oceans are becoming more acid, and the more acid an ocean is the less CO_2 it can absorb.

Before the end of this century the oceans are predicted to be taking in 10 per cent less CO_2 than they do today. In the meantime we continue to pour more and more CO_2 into the atmosphere.

FOUR

ICE AGES AND SUNSPOTS

Why doesn't the Earth retain all the heat it receives from the Sun? On the other hand, why doesn't all the heat escape again back into space?

Think of what happens when you visit a ski field and the air remains cold on a sunny day. This is because the Sun does not warm the atmosphere (and there is little water vapour in the cold air to trap any heat) and because its energy is reflected back into space by the snow. But when its rays fall on a darker surface, such as skin or a ski glove, the rays are absorbed and heat is generated.

As your ski glove becomes toasty, the heat energy is radiated back into the sky, where it is captured by the greenhouse gases in the atmosphere. And so light passes harmlessly through an atmosphere charged with greenhouse gases, but heat has trouble getting out.

Many scientists have asked questions about what causes the Earth to heat and cool. One of the most remarkable was Milutin Milankovich, who spent most of his career working as a civil engineer in the Austro-Hungarian Empire. Born in

1879, in what is now Serbia, he was interned during World War I in Budapest, where he was allowed to work in the library of the Hungarian Academy of Sciences. He had already begun to ponder on that great puzzle of his times—the cause of the ice ages. More than two decades later, in 1941, with the world embroiled in yet another global conflict, he was finally ready to publish his great work, *Canon of Insolation of the Ice-Age Problem.*

Milankovich identified three principal cycles that drive Earth's climatic variability. The longest of the cycles concerns the planet's orbit around the Sun. Surprisingly perhaps, Earth's orbit is not a perfect circle but an ellipse whose shape changes on a 100,000-year cycle known as Earth's eccentricity. When Earth's orbit is strongly elliptical, the planet is carried both closer to and further away from the Sun, meaning that the intensity of the Sun's rays reaching the Earth varies through the year.

At present the orbit is not very elliptical, and there is only a 6 per cent difference between January and July in the radiation reaching Earth. At other times that difference is 20 to 30 per cent. This is the only cycle that changes the total amount of the Sun's energy reaching Earth, so its influence is considerable.

The second cycle takes 42,000 years to run its course, and it concerns the tilt of Earth on its axis. This varies from 21.8 to 24.4 degrees, and it determines where the most radiation will fall. At the moment the Earth's axial tilt is in the middle of its range.

The third and shortest cycle, which runs its course every 22,000 years, concerns the wobble of Earth on its axis. During the course of this cycle, Earth's axis shifts from pointing to the Pole Star to pointing towards Vega. This affects the intensity of the seasons. When Vega marks true north, winters can be bitterly cold and summers scorchingly hot.

So when do Milankovich's cycles cause ice ages?

The answer is connected with the way the continents drift around the surface of the Earth. When continental drift brings large parts of Earth's land surface near the Poles, and when the cycles are right, mild summers and harsh winters allow snow to accumulate on the polar lands. Eventually the snow builds into great ice domes and an ice age is born.

Even at their most extreme, Milankovich's cycles bring an annual variation in the total amount of sunlight reaching Earth of less than one tenth of 1 per cent. Yet that seemingly trivial difference can cause Earth's temperature to rise or fall by a whopping 5°C.

Exactly how remains a profound mystery, but it is certain that greenhouse gases play a role. Indeed, computer models cannot simulate the onset of an ice age unless atmospheric CO_2 is reduced in the Southern Hemisphere.

Milutin Milankovich solved the riddle of the ice ages, but it was decades before the world learned of his brilliance.

His *Canon* was translated into English in 1969. By then sediments drawn from the deep ocean floor had given oceanographers direct evidence of exactly the kind of impacts he had predicted.

These studies revealed that Milankovich cycles should be cooling Earth. In the early 1970s, when this became widely understood, scientists began to talk of a new ice age, but that was before they realised how human pollution was altering the balance of greenhouse gases.

Today, Milankovich's masterwork is regarded as one of the greatest breakthroughs ever made in climate studies.

With an understanding of greenhouse gases and Milankovich cycles, climatologists were on their way to grasping why Earth's climate has varied over time. Yet there are still other factors to consider.

One is the intensity of radiation emitted from the Sun. About two-thirds of the Sun's rays reaching our planet are absorbed and put to work here, while the remaining third is reflected back into space.

Over 2000 years ago Greek and Chinese astronomers wrote of seeing dark spots on the Sun whose shape and location changed. In April 1612 the Italian astronomer Galileo, armed with one of the first telescopes, made detailed observations of these sunspots. He showed that they were not satellites passing across the surface of the Sun, but originate from the star itself.

In the nineteenth century it was discovered that the activity of sunspots varies on an eleven-year cycle, as well as on a longer cycle of centuries. Sunspots are slightly cooler than the rest of the Sun's surface, yet oddly enough when there are lots of them Earth seems to warm up. A scarcity of sunspots is thought to account for around 40 per cent of the temperature decrease experienced during the so-called Maunder Minimum of 1645–1715 when temperatures in Europe dropped by around a degree.

Do sunspots affect Earth's climate as a whole? A recent study of tree-rings going back 6000 years failed to find any evidence that sunspot activity affects tree growth. So while sunspots undeniably exist, their impact on Earth's living organisms (and thus atmosphere) must be too small to measure.

Scientists recently discovered that variations in solar radiation and greenhouse gas concentrations affect Earth's climate in different ways. Solar radiation warms the upper levels of the stratosphere through the ultraviolet rays that are absorbed by ozone. Greenhouse gases, in contrast, warm the troposphere, and they warm it most at the bottom where their concentration is greatest.

At the moment Earth is experiencing both stratospheric cooling (due to the ozone hole) and tropospheric warming (due to increased greenhouse gases). Sunspots cannot be responsible for this.

The fossil record can also teach us much about climate. It is characterised by sudden shifts from one steady climatic state to another. It is as if our planet reacts in jolts to the factors that influence climate. This series of wild shifts has in the past driven animals and plants from one end of a continent to another.

FIVE

TIME'S GATEWAYS

Geology students who have to memorise the divisions of the geological time scale often resort to naughty nonsense phrases like 'Can Ollie See Down Mike's Pants' Pockets? Tom Jones Can. Tom's Queer.' The C in 'Can' is for Cambrian, O in 'Ollie' is for Ordovician, S in 'See' is for Silurian, and so on through to our own time, the Quaternary.

Having memorised this list, students have only learned the basics, for each major division is itself divided into Periods, which are in turn divided into local units. These finest divisions of time are called local units because they are only recognised in limited areas. In North America, for example, the Periods of the Cainozoic Era are divided into local units known as 'North American land mammal ages'. Although they are the smallest divisions on the time scale, many lasted for several million years.

The divisions of the geological time scale can easily be told apart, because of what geologists call 'faunal turnover'—times when species suddenly appear or disappear.

We can think of these episodes as time's gateways—occasions when one Era, and often one climate, gives way to the next.

There are just three agents of change sufficiently powerful to open a gateway in time—the shifting of continents, cosmic collisions, and climate-driving forces such as greenhouse gases. All act in different ways, but they drive evolution using the same mechanisms—death and opportunity.

Time's gateways come in three 'sizes'—small, medium and large. A small gateway might occur when continents bump into each other, or when land-bridges form as seas rise and fall, or when Earth heats and cools. In these instances, time's gateways are marked by the sudden arrival of new species, and often the extinction of local competitors.

The medium-sized divisions of time—those separating geological Periods—are global in scale. In these cases, what you read in the rocks is almost invariably a sorry tale of extinction followed by the slow evolution of new life forms that adapt to the changed conditions.

Time's greatest divisions, however, are those separating Eras. These are occasions of massive upheaval, when as much as 95 per cent of all species vanish. Our planet has experienced such massive extinctions on only five previous occasions.

The last time Earth was afflicted was 65 million years

ERA	PERIOD	EPOCH	SIGNIFICANT EVENT	YEARS AGO
				present day
Cainozoic	Quaternary	Holocene	The Long Summer	8000
		Pleistocene	Ice Ages *First modern humans*	1.8 million
	Tertiary	Pliocene	*First upright human ancestors*	5.3 million
		Miocene	*Decline of widespread rainforests*	23.8 million
		Oligocene	*Diverse vertebrate communities*	33.7 million
		Eocene	*Final separation of Australia from Antarctica* Clathrate release 55 million years ago	55.5 million
		Palaeocene	Cretaceous–Tertiary extinction about 65 million years ago	65 million
Mesozoic	Cretaceous		*First flowering plants*	145 million
	Jurassic		*First birds*	213 million
	Triassic		*First dinosaurs* Permian–Triassic extinction about 251 million years ago	248 million
Palaeozoic	Permian		*First conifers; early reptiles* Ice Ages about 350 to 250 million years ago	286 million
	Carboniferous		*Early amphibians* Late Devonian extinction about 364 million years ago	360 million
	Devonian		*First insects*	410 million
	Silurian		*First fish* Ordovician–Silurian extinction about 439 million years ago	440 million
	Ordovician		*Marine invertebrates*	505 million
	Cambrian		Cambrian explosion	544 million
Proterozoic			Ice Ages about 800 to 600 million years ago	2500 million
Achaean			*First life*	3800 million
Hadean			*Earth takes form*	4500 million

Geological Time

ago, when every living thing weighing more than 35 kilograms, and a vast number of smaller species, was destroyed. This is when the dinosaurs vanished, and the cause is widely believed to have been an asteroid colliding with Earth. So much debris exploded into the atmosphere that the climate changed, which caused the great global dying.

CO_2, it turns out, played a major role in the event. By studying fossil leaves, palaeobotanists know that atmospheric CO_2 grew massively after the impact, probably because the asteroid collided with limestone-rich rock. This instantaneous injection of greenhouse gas would have caused an abrupt spike in temperature. Species that could not cope with the increased heat (including many reptiles) would have perished.

Ten million years later—55 million years ago—there was another global event. The Earth's surface abruptly heated by 5 to 10°C. In November 2003 scientists drilling over two kilometres below the floor of the north Pacific Ocean, encountered a 25-centimetre-thick layer of ooze. Its analysis revealed an astonishing tale.

The first thing the researchers noted was that the layer sat above a section of sea floor that had been eaten away by acid, powerful proof that the oceans had turned acidic. It's a trend that we can observe today and which occurs when CO_2 is being absorbed by sea water in large amounts.

Not surprisingly, life in the ocean depths was profoundly

affected. By studying fossils, researchers could work out that there were massive extinctions of marine life, from the tiny plankton to the monsters of the deep.

On land there is evidence for abrupt changes in rainfall during this period. And there was a remarkable series of migrations in which the fauna and flora of Asia streamed across land bridges that ran through the Arctic Circle into North America and Europe. The new arrivals drove many creatures to extinction.

We now know that back then a mind-boggling 1500 to 3000 gigatonnes of carbon was injected into the atmosphere. From a geological perspective the release happened 'instantaneously', meaning it may have occurred over decades or less. Atmospheric concentrations of CO_2 rose from around 500 parts per million (twice the concentration of the last 10,000 years) to around 2000 parts per million.

The climate change of 55 million years ago seems to have been driven by a vast, natural gas-driven equivalent of a barbecue.

Scientists believe that the gas may have come from craters below the sea off the Norwegian coast. The fuel for the event lay in one of the greatest accumulations of hydrocarbons—mostly in the form of methane gas—that we know about.

We can imagine Earth's crust creaking as the hot tongues of molten rock made their way towards the fuel. Most probably it did not burn, but heated and expanded, forcing its way quickly towards the surface. When it arrived on the sea floor a massive submarine explosion must have ensued, the likes of which the world had never seen.

Most of the methane, however, did not reach the atmosphere. Instead it combined with oxygen in the sea water, leaving only CO_2 to arrive at the surface. With the deep ocean devoid of oxygen, life must have struggled. Then, as CO_2 turned the depths acidic, a cavalcade of creatures, most of which will never be known to us, were force-marched off to extinction. Indeed there is mounting evidence that many of the deep-sea creatures that are with us today evolved after this time. It took at least 20,000 years for the Earth to re-absorb all of the additional carbon.

Because the extinction event of 55 million years ago was caused by a rapid increase in greenhouse gases, it offers the best parallel with our current situation. Yet there are significant differences too.

Earth has now been in an icehouse phase for millions of years, but 55 million years ago it was already very warm, with CO_2 levels around twice the level they are today. There were no ice caps then, and presumably fewer cold-adapted species—certainly nothing like narwhals and polar bears. Nor was this warmer world likely to have possessed the wondrous slices of life we find today on mountains and in the depths of the sea.

Now Earth stands to lose far more from rapid warming than the world of 55 million years ago. Back then the warming closed a geological Period, while we might, through our activities, bring to an end an entire Era.

SIX

BORN IN THE DEEP-FREEZE

We human beings are, as our scientific name *Homo sapiens* suggests, the 'thinking creatures'. We are also, in the grand scheme of things, very recent arrivals.

The Epoch that gave birth to us is called the Pleistocene, a word that means the most recent times. It covers the last 2.4 million years. The first of our kind—moderns in every physical and mental respect—strode about Earth around 150,000 years ago in Africa, where archaeologists have found bones, tools and the remains of ancient meals. These people had evolved from small-brained ancestors known as *Homo erectus*, who had been in existence for nearly 2 million years.

Perhaps the driving force that changed some of 'them' into 'us' was the opportunity offered by the rich shorelines of the African rift lakes, or perhaps the bounty of the Agulhas Current that runs along the continent's southern shores. In such places new foods and challenges may have favoured specialised tool use and given an evolutionary advantage to people of high intelligence.

The environment of these distant ancestors was

dominated by an icehouse climate in which the fate of all living things was determined by Milankovich's cycles. Whenever these cycles expanded the frozen world of the Poles, chill winds blew all over the planet and temperatures plummeted. Lakes shrank or filled, bountiful sea-currents flowed or slackened, and vegetation and animals alike undertook continent-long migrations.

The genetic inheritance laid down in this world of ice is still with us. A great reduction in the diversity of our genes, for example, tells of a time around 100,000 years ago when humans were as rare as gorillas are today. We could so easily have vanished, for 2000 fertile adults were all that stood between us and extinction.

But then the Milankovich cycles altered in ways that favoured our species, and by 60,000 years ago small bands of humans had wandered across the Sinai and out into Europe and Asia. By 46,000 years ago they had reached the island continent of Australia, and by 13,000 years ago, as the ice waned for a final time, they discovered the Americas.

Now there were millions of us on the planet, and groups thrived from Tasmania to Alaska. Yet for thousands of years these intelligent people, who were like us in every physical and mental way, remained nothing but hunters and gatherers. In the light of our great accomplishments over the past 10,000 years, this long period without significant cultural development is a riddle. Is the riddle to do with climate? To answer that, we need to look at the ice-age climate record.

One source of information about climate is any piece of timber. You can see in its growth rings the story of how things were when that tree lived.

Widely spaced rings tell of warm and bountiful growing seasons when the sun shone and rain fell at the right time. Compressed rings, recording little growth in the tree, tell of adversity when long, hard winters or drought-blighted summers tested life to the limits.

The oldest living thing on our planet is a bristlecone pine growing more than 3000 metres up in the White Mountains of California. More than 4600 years old, it survives in Methuselah Grove alongside many other superannuated specimens. Its precise location is a closely guarded secret because, vulnerable to disturbance, it's been slowly dying for the past 2000 years.

Within its trunk this single tree holds a detailed, year-by-year record of climatic conditions in California. Match the pattern from the heart of the Methuselah tree with the rind of a dead stump nearby, and you may pierce time to a depth of 10,000 years. Tree-ring records of this length have now been obtained from both hemispheres. There is even hope that the great kauri pines of New Zealand, whose timber can lie in swamps for millennia without rotting, will provide a record spanning 60,000 years of climatic change.

For all its convenience, the climate record of the trees is relatively limited in what it can tell us. If you want a really detailed record you must turn to ice—but only in special places does it yield all its secrets.

One such place is the Quelccaya ice cap in the high mountains of Peru. Each year's snowfall there is separated by a band of dark dust blown up from the deserts below during the winter dry season. Three metres of snow can fall on Quelccaya in a summer, and the falls of subsequent seasons compress it, first to firn (compacted snow) and then to ice.

In the process, bubbles of air are trapped, which become minute archives documenting the condition of the atmosphere. Even the dust in the bubbles is informative, for it tells of the strength and direction of the winds, and of conditions below the ice cap.

The ice sheets of Greenland and Antarctica yield Earth's longest cores. When the circumstances are right, spectacular records can be extracted. In June 2004, an ice-core over three kilometres long was drawn from a region of the Antarctic known as Dome C (about 500 kilometres from the Russian Vostok base). Drilling through ice is more hazardous than you might imagine, and the recovery of such a long ice-core must count as one of science's greatest triumphs.

The drill site was bitterly cold: –50°C at the beginning of the drilling season and –25°C in the middle of the Antarctic summer. The drill itself is just ten centimetres wide, and as it grinds its way downward a slender column of ice is separated and drawn to the surface. The first kilometre was especially difficult, for there the ice is packed with air bubbles. As the core was drawn up these tended to depressurise, shattering the ice into useless shards. Worse, ice chips can clog the drill head, jamming it fast.

In the summer of 1998–99 a drill head trapped over a kilometre below the surface forced the abandonment of the hole, leaving the team with no option but to start all over again. This time, as they drilled the three kilometres to the bottom, they stopped after each metre or two to bring the precious core to the surface.

As the team passed the point reached by earlier drilling, the excitement was palpable. 'You knew you were getting stuff that had never been seen before,' a team member said, and each kilometre advanced was celebrated with specially warmed champagne.

When they were almost at bedrock, another problem emerged. Heat from the rocks below was softening the ice, threatening yet another jamming of the drill bit. The final hundred metres were drilled in late 2004, using as a makeshift bit a plastic bag filled with ethanol (to melt its way downwards).

The core from Dome C allows us to glimpse conditions during the so-called glacial maxima, when the grip of the ice is at its greatest. The last time this happened was between 35,000 and 20,000 years ago.

Back then the sea level was more than 100 metres lower

than it is today, altering the very shape of the continents. North America and Europe's most densely inhabited landscapes lay under kilometres of ice. Even regions south of the ice, such as central France, were treeless subarctic deserts. Their growing season of sixty days was an alternation of freezing northerly winds and a few still periods when a stifling haze of glacial dust filled the air.

By the end of the ice age, changes were big and moving very fast indeed. Climatologists are especially interested in the period from around 20,000 to 10,000 years ago—as the glacial maximum began to wane—for over those ten millennia the overall surface temperature of Earth warmed by 5°C—the fastest rise recorded in recent Earth history.

How does the rate and scale of change during this period compare with what is predicted to happen this century? If we do not reduce our emissions of greenhouse gases, an increase of 3°C (give or take 2°C) over the twenty-first century seems inevitable. But at the end of the last glacial maximum, the fastest warming recorded was a mere 1°C per thousand years.

Today we face a rate of change thirty times faster—and, because living things need time to adjust, speed is every bit as important as scale when it comes to climate change.

In 2000, analysis of a core from Bonaparte Gulf in Australia's tropical northwest revealed that 19,000 years ago, over a period of just 100 to 500 years, sea levels rose abruptly by ten to fifteen metres, indicating that the thaw commenced far earlier than anyone had imagined. The water came from the collapse of a massive Northern Hemisphere ice sheet. Its meltwater in turn had an impact on the Gulf Stream, the extraordinary ocean current that flows north for thousands of kilometres from the Gulf of Mexico.

Back then, the melting ice sheet poured somewhere between one quarter and two Sverdrups' worth of fresh water into the North Atlantic. The scale of ocean currents is measured in Sverdrups, named after the Norwegian oceanographer Hans Ulrich Sverdrup. A Sverdrup is a very large flow of water—1 million cubic metres of water per second—and by disrupting the Gulf Stream this influx had profound consequences.

The Gulf Stream transports vast amounts of heat northward from near the equator—almost a third as much as the Sun brings to Western Europe, and that heat is borne in a stream of warm salty water. As it gives up its heat the water sinks because, being salty, it is heavier than the water around it. This sinking draws more warm, salty water northwards. If the Gulf Stream's saltiness is diluted with fresh water it does not sink as it cools, and no more warm water is drawn northward in its wake.

The Gulf Stream has stopped flowing altogether in the past. Without the heat it brings, the melting glaciers begin to grow again and, as their white surface reflects the Sun's heat back to space, the land cools. Animals and plants migrate or die and temperate regions such as central France are plunged into a Siberian chill.

The heat, however, does not vanish. Most of it pools around the equator and in the Southern Hemisphere, where it can cause the melting of glaciers in the south. The Sun's rays then fall on a dark sea surface instead of on ice, and are absorbed. This heats the world from the bottom up, so to speak. With the Gulf Stream established once more, courtesy of growing northern ice, the planet enters another cycle of warming.

Somewhere around two Sverdrups of fresh water is required to slow the Gulf Stream, and the geological record confirms that this happened repeatedly between 20,000 and 8000 years ago. Thus the transition from the ice age to the warmth of today was the wildest of roller-coaster rides.

And then this climatic madness gave way to the most serene calm. It was as if, says archaeologist Brian Fagan, a long summer had arrived whose warmth and stability the ice-age world had not seen before.

All over the world people who had been sheltering in huts and living hand to mouth began to grow crops, domesticate animals and live in settled towns.

Were warmth and stability the triggers that caused the flowering of our complex society?

SEVEN

THE LONG SUMMER

The long summer of the last 8000 years is without doubt *the* crucial event in human history. Although agriculture commenced earlier (around 10,500 years ago in the Fertile Crescent in the Middle East), it was during this period that we acquired most of our major crops and domestic animals, the first cities came into being, the first irrigation ditches were dug, the first words written down, and the first coins minted.

And these changes happened not once, but independently many times in different parts of the world. Before our long summer was 5000 years old, cities had sprung up in Western Asia, East Asia, Africa and central America. The similarities in their temples, homes and fortifications are astonishing.

It is as if the human mind contained a plan for the city all along, and was just waiting until conditions permitted to build it.

These human settlements were ruled by an elite who relied on artisans. In a few societies writing developed, and in even the earliest of these jottings—clay tablets from ancient Mesopotamia—we recognise life as it is lived in a great metropolis.

Did this long summer result from a cosmic fluke? Were Milankovich's cycles, the Sun and Earth all 'just right' to create a warm period of unprecedented stability? During every warm period we know about over the last million years Milankovich's cycles caused a sudden spike in temperature followed by a long, unstable cooling. There is nothing unique about the current Milankovich cycle that can account for the long summer. Indeed, were Milankovich cycles still controlling Earth's climate, we should be feeling a distinct chill by now.

As he tried to explain the long summer, Bill Ruddiman, an environmental scientist at the University of Virginia, began to look for a unique factor—something that was operating only in this last cycle, but in none of the earlier ones. That unique factor, he decided, was us.

The Nobel laureate Paul Crutzen (awarded the prize for research into the ozone hole) and his colleagues had already recognised and named a new geological Period in honour of our species. They called it the Anthropocene—meaning the age of humanity—and they marked its dawn at 1800 when methane and CO_2, brewed up by the gargantuan machines of the Industrial Revolution, first began to influence Earth's climate.

Ruddiman added a revolutionary twist to this argument:

he detected what he believes to be human influences on Earth's climate that occurred long before 1800.

Charting the levels of two critical greenhouse gases—methane and CO_2—in air bubbles trapped in the Greenland and Antarctic ice sheets, Ruddiman discovered that, beginning 8000 years ago, the Milankovich cycles could not explain what actually happened. Methane should have commenced declining at that time, and gone into a rapid decline by 5000 years ago. Instead, after taking a shallow dip, methane concentrations begin a slow but emphatic rise.

This, Ruddiman argues, is evidence that humans had wrested control of methane emissions from nature, and so we should mark the dawn of the Anthropocene at 8000 years ago rather than 200.

It was the beginnings of agriculture—particularly wet agriculture of the kind practised in flooded rice paddies in eastern Asia—that tipped the balance. These agricultural systems can be prodigious producers of methane. Farmers of other crops that require swampy conditions were making their own contributions at around this time. Taro agriculture (which involves the creation and maintenance of water-controlling structures), for example, was well under way in New Guinea by 8000 years ago.

Even hunter-gatherers may have had a role. The construction of weirs transformed vast areas of southeastern Australia into seasonal swamps. These structures were perhaps the most extensive ever created by non-agricultural people, and were used to regulate swamps for the production of eels. Harvested en masse at great gatherings of the tribes, the eels were then dried and smoked to be traded over large distances.

Ruddiman also found evidence in the ice bubbles that the concentration of CO_2 in the atmosphere was being influenced by humans far earlier than first imagined. CO_2 levels rise rapidly as the glacial stage ends, then typically begin a slow decline towards the next cold period. But in this cycle they kept rising. By 1800 atmospheric CO_2 had risen to 280 parts per million. If natural cycles were still solely in control of Earth's carbon budget, Ruddiman states, CO_2 should have then stood at only 240 parts per million, and be declining.

At first glance his argument looks flimsy. After all, early humans would have needed to emit twice as much carbon as our industrial age did between 1850 and 1990—an output only made possible by an unprecedented population using coal-burning machines.

The key, notes Ruddiman, is time. Eight thousand years is a long span, and as humans cut and burned forests around the globe their activities acted like a hand casting feathers on a set of scales: eventually enough feathers piled up to tip the balance.

The delicate climatic stability created by humanity over

the past 8000 years, Ruddiman argues, was still vulnerable to the great cycles of Milankovich. The archaeologist Brian Fagan argues that these cycles could be amplified into truly monumental impacts on human societies. The slight shift in Earth's orbit between 10,000 and 4000 BC brought between 7 and 8 per cent more sunlight to the Northern Hemisphere.

This changed atmospheric circulation, which resulted in increased rainfall in Mesopotamia by 25 to 30 per cent. What was once a desert was transformed into a verdant plain that supported dense farming communities. After 3800 BC, however, Earth's orbit reverted to its former pattern and rainfall dropped off, forcing many farmers to abandon their fields and wander in search of food.

Fagan believes that the famine-driven wanderers found refuge in a few strategic locations such as Uruk (now in southern Iraq), where irrigation canals branched off the main rivers. Here the starving migrants were put to work by a central authority in construction projects such as the maintenance of canals.

Reduced rainfall, Fagan argues, also forced Uruk's farmers to innovate, and so they used, for the first time, ploughs and animals to till fields in a rotation that involved producing two crops per year.

With grain production localised around strategic towns, surrounding settlements began to specialise in producing goods such as pottery, metals or fish, which were traded at

Uruk's markets for the ever-scarcer grain.

Each of these changes led to the development of a more centralised authority, which in turn employed the world's first bureaucrats, whose job it was to tally and distribute the vital grain.

The sum of all of this change was a shift in human organisation, and by 3100 BC Mesopotamia's southern cities had become the world's first civilisations. Indeed the city, Fagan argues, is a key human adaptation to drier climatic conditions.

Let's return now to Bill Ruddiman's analysis, because it contains several twists in its tail. He sees a clear correlation with times of low atmospheric CO_2 and several plagues caused by the bacterium *Yersinia pestis*—the black plague of medieval times. These epidemics were global in their reach and killed so many people that forests were able to grow back on deserted farmland. In the process they absorbed CO_2, lowering atmospheric concentrations by 5 to 10 parts per million. Global temperatures then fell and periods of relative cold ensued in places such as Europe.

Ruddiman's thesis implies that, by adding sufficient greenhouse gases to delay another ice age, yet not overheating the planet, the ancients performed an act of chemical wizardry. Today, however, the changes scientists are detecting in our atmosphere are so great that time's gates appear once again to be opening.

There are unmistakable signs that the Anthropocene is turning ugly. Will it become the shortest geological Period on record?

EIGHT

DIGGING UP THE DEAD

We walk on earth,
we look after,
like rainbow sitting on top.
But something underneath,
under the ground.
We don't know.
You don't know.
What do you want to do?
If you touch,
you might get cyclone, heavy rain or flood.
Not just here,
you might kill someone in another place.
Might be kill him in another country.
You cannot touch him.

Big Bill Neidjie, *Gagadju Man*, 2001.

Australia's Aborigines live close to the land, and they have a distinctive way of viewing the world. They tend to see a whole picture. Big Bill Neidjie was a truly wise elder who spent his youth living a tribal life always moving, hunting and gathering. When he tells us about the impact of mining in his Kakadu country he doesn't talk of the mines, the tailings and the poisoned earth. In just a handful of words he describes the great cycle that runs from disturbing the eternal living dreaming of the ancestors to the catastrophe awaiting unborn generations.

The challenge he throws down—'What do you want to do?'—is discomforting, because my country—Bill's country—is pierced through and through with mines of every type, and more coal is drawn from its innards to be shipped off overseas than from any other place on the planet. Yet Bill has intuited the hidden links between mining, climate change and human wellbeing that scientists struggle to understand in their studies of the greenhouse effect. Bill's challenge remains there to be answered, because we still have a chance to decide our future.

Fossil fuels—oil, coal and gas—are all that remain of organisms that, many millions of years ago, drew carbon from the atmosphere. When we burn wood we release carbon that has been out of atmospheric circulation for a few decades, but when we burn fossil fuels we release carbon that has been out of circulation for eons.

Digging up the dead in this way is a particularly bad thing for the living to do.

In 2002, the burning of fossil fuels released a total of 21 billion tonnes of CO_2 into the atmosphere. Of this, coal contributed 41 per cent, oil 39 per cent, and gas 20 per cent.

The energy we liberate when we burn these fuels comes from carbon and hydrogen. Because carbon causes climate change, the more carbon-rich a fuel is, the greater danger it presents to humanity's future. Anthracite, the best black coal, is almost pure carbon. Burn a tonne of it and you create 3.7 tonnes of CO_2.

The fuels derived from oil contain two hydrogen atoms for every one of carbon in their structure. Hydrogen gives more heat when burned than carbon (and in doing so produces only water), so burning oil releases less CO_2 than coal does.

The fossil fuel with the least carbon content is methane, which has just one carbon atom for every four hydrogen atoms.

These fuels form a stairway leading away from carbon as the fuel for our economy. Even using very advanced methods (and most coal-fired power plants come nowhere near this), burning anthracite to generate electricity results in 67 per cent more CO_2 emissions than does methane, while brown coal (which is younger, and has more moisture and impurities) produces 130 per cent more.

To see the most polluting power plant operating in the world's major industrialised countries (in terms of CO_2), you need to travel to Gippsland, Victoria, where the brown-coal fired Hazelwood power station provides electricity for much of the state.

From a climate change perspective, then, there's a world of difference between burning gas or coal.

Coal is our planet's most abundant fossil fuel. Those in the industry often refer to it as 'buried sunshine', because coal is the fossilised remains of plants that grew in swamps millions of years ago. In places like Borneo you can see the initial stages of the coal-forming process taking place. There, huge trees topple over and sink into the quagmire, where a lack of oxygen impedes rotting. More and more dead vegetation builds up until a thick layer of sodden plant matter is in place. Rivers then wash sand and silt into the swamp, which compresses the vegetation, driving out moisture and other impurities.

As the swamp is buried deeper and deeper in the earth, heat and time alter the chemistry of the wood, leaves and other organic matter to produce decomposing debris called peat. First, the peat is converted to brown coal and, after many millions of years, the brown coal becomes bituminous coal, which has less moisture and impurities. If further pressure and heat are applied, and more impurities removed, it can become anthracite. At its most exquisite anthracite—in the form of jet—is a jewel as beautiful as a diamond.

Certain times in Earth's history have been better for forming coal than others. In the Eocene Period, around 50 million years ago, great swamps lay over parts of Europe and Australia. Their buried remains form the brown coal deposits found today.

Much of the world's anthracite lived during the Carboniferous Period, 360 to 290 million years ago. Named for the immense coal deposits that were then laid down, the Carboniferous world was a very different place from the wetlands of today.

If you'd been able to punt through the ancient swamps of that bygone era, instead of river red gums or swamp cypress, you would have seen gigantic relatives of clubmosses and lycopods (primitive swamp plants), as well as far stranger plants that are now extinct. The scaly, columnar trunks of the *Lepidodendron* grew in dense forests, each trunk two metres in diameter and soaring forty-five metres into the air. They did not branch until near their tips, where a few short, straggly sproutings bore metre-long grassy leaves.

There were no reptiles, mammals or birds in those long ago times. Instead, the stifling, humid growth teemed with insects and their kin. The atmosphere was rich in oxygen. Millipedes grew to two metres in length, spiders reached a metre across, and cockroaches were thirty centimetres long. There were dragonflies whose wingspans approached a metre, while in the waters below lurked crocodile-sized amphibians with huge heads, wide mouths and beady eyes.

In stealing the buried treasure of this alien world we have set ourselves free from the limits of biological production in our own era.

The march towards a fossil-fuel-dependent future started in the England of Edward I. The King himself so detested the smell of coal that in 1306 he banned the burning of it. There are even records of coal burners being tortured, hanged or decapitated. But England's forests were becoming exhausted. In spite of the King, the English became the first Europeans to burn coal on a large scale.

People had no idea what coal was. Many miners believed that it was a living substance that grew underground, and would grow faster if it were smeared with dung. The stench of brimstone (sulphur) that accompanied its burning was a reminder of the torments of hell underground. People also associated coal with the plague.

Despite all this, by 1700 a thousand tonnes per day was being burned in the city of London. An energy crisis soon loomed. England's mines had been dug so deep that they were filling with water. A way of pumping it out had to be found.

The man who discovered how this might be done was a small-town ironmonger, Thomas Newcomen. His device burned coal to produce steam, which was then condensed to create a vacuum that moved a piston which pumped the water. The first Newcomen engine was installed in a Staffordshire coal mine in 1712. Fifty years later, hundreds of them were at work in mines across the nation, and England's coal production had grown to 6 million tonnes per year.

The ingenious James Watt improved on Newcomen's design and, in 1784, Watt's friend William Murdoch produced the first mobile steam engine. From that moment on, the coming century—the nineteenth—was to be the century of coal. No other fuel source could rival it for cooking, heating, industrial purposes and transport. In 1882 when Thomas Edison opened up the world's first electric light power station in lower Manhattan, electricity production was added to coal's portfolio.

More coal is burned today than at any time in the past.

Two hundred and forty-nine coal-fired power plants were projected to be built worldwide between 1999 and 2009, almost half in China. A further 483 will follow in the decade to 2019, and 710 more between 2020 and 2030. About a third of these will be Chinese, and in total they will produce 710 gigawatts of power. A gigawatt is a billion watts of power. The CO_2 they produce will continue to warm the planet for centuries.

If the nineteenth was the century of coal, the twentieth was the century of oil. On 10 January 1901, on a small hill in Texas called Spindletop, Al Hamill was drilling for oil. He had penetrated more than 300 metres into the sandstone below, and by 10.30 in the morning, disgusted by his lack of success, was about to give up. At that moment, 'with a deafening blast, and a great howling roar, thick clouds of methane gas jetted from the hole. Then came the liquid, a column of it, six inches wide. It rocketed hundreds of feet into the winter sky before falling back to earth as a dark rain.' The discovery of oil in such deep strata was new. Oil drove coal from the fields of transport and home heating.

The trouble with oil is that there is far less of it than coal and it's harder to find.

Oil is the product of life in ancient oceans and estuaries. It is composed primarily of the remains of plankton—in particular single-celled plants known as phytoplankton. When the plankton die their remains are carried down to oxygen-free depths, where their organic matter can accumulate without being consumed by bacteria.

The geological process for making oil is as precise as a recipe for making pancakes.

First the sediments containing the phytoplankton must be buried and compressed by other rocks. Then, the absolutely right conditions are needed to squeeze the organic matter out of the source rocks and to transfer it, through cracks and crevices, into a suitable storage layer. This layer must be porous, but above it must lie another rock thick enough to forbid escape.

In addition, the waxes and fats that are the source of oil need to be 'cooked' at between 100–135°C for millions of years. If the temperature exceeds these limits, all that will result is gas, or else the fossil fuel will be lost entirely. The creation of oil reserves is the result of pure chance—the right rocks being cooked in the right way for the correct time.

The house of Saud, the Sultan of Qatar and the other wealthy principalities of the Middle East all owe their good fortune to this geological accident. The conditions in the rocks of their region have been just right to deliver a bonanza of oil. Before it was tapped, just one Saudi Arabian oilfield, the Ghawar, held a seventh of the entire planet's oil reserves.

Until 1961 the world's oil companies were finding more and more oil every year, much of it in the Middle East. Since then the rate of discovery has dwindled, yet rates of use have gone up. By 1995 humans were using an average of 24 billion barrels of oil per year, but an average of only 9.6 billion barrels was discovered. By 2006 oil was over $US70 a barrel. Many analysts are predicting even higher prices, and perhaps shortages, as soon as 2010, which suggests that something else will be needed to power the economies of the twenty-first century.

That 'something else', many in the industry believe, is natural gas, around 90 per cent of which is methane. Thirty years ago gas supplied just 20 per cent of the world's fossil fuel. If current trends continue, by 2025 it will have overtaken oil as the world's most important fuel source. There are proven reserves of gas sufficient to last fifty years. It's looking likely that this will be the century of gas.

In 1900 the world was home to a billion people or so.

By 2000 there were 6 billion of us, each using on average four times as much energy as our forefathers did 100 years earlier. In the twentieth century the burning of fossil fuels increased sixteen-fold.

As the researcher Jeffrey Dukes says, all the carbon and hydrogen in fossil fuels was gathered together through the power of sunlight, captured by long-ago plants. He has calculated that approximately 100 tonnes of ancient plant life is required to create four litres of petrol.

This means that, over each year of our industrial age, humans have required several centuries' worth of ancient sunlight to keep the economy going. The figure for 1997—around 422 years of fossil sunlight—was typical.

More than 400 years of blazing light from the Sun—and we burn it in a single year!

Dukes's ideas have changed the way I look at the world. Now, as I tread the sandstone pavements around Sydney, I feel the power of long-spent sunbeams on my bare feet. Looking at the rock through a magnifying lens I can see the grains whose rounded edges caress my toes, and I realise that each one of the countless billion grains has been shaped by the power of the Sun. Over 300 million years ago that Sun drew water from an ocean which then fell as rain on a distant mountain range. Bit by bit the rock crumbled and was carried into streams, until all that remained were rounded grains of quartz.

A million times more energy must have gone into creating sand grains than has ever gone into all human enterprise. From the soles of my feet to the top of my sun-warmed head, I instantly know what Dukes is saying about fossil sunlight: the past is an abundant land, whose buried riches are fabulous when compared with our daily ration of solar radiation.

The power and seduction of fossil fuels will be hard to leave behind. If humans were to look to biomass (all living things, but in this case particularly plants) as a replacement, we would consume 50 per cent more than we now produce on land. We're already using 20 per cent more than the planet can sustainably provide, so we need to find sustainable, innovative ways to do this.

In 1961 there were just 3 billion people, and they were using half of the total resources that our global ecosystem could sustainably provide. By 1986, our population topped 5 billion, and we were using *all* of Earth's sustainable production.

By 2050, when the population is expected to level out at around 9 billion, we will be using—if they can still be found—nearly two planets' worth of resources. But for all the difficulty we'll experience in finding those resources, it's our waste—particularly the greenhouse gases—that is the limiting factor.

Since the beginning of the Industrial Revolution a global warming of 0.63°C has occurred on our planet. Its principal cause is an increase in atmospheric CO_2 from around three parts per 10,000 to just under four. Most of the increase in the burning of fossil fuels has occurred over the last few decades.

Nine out of the ten warmest years ever recorded have occurred since 1990.

2

ONE IN TEN THOUSAND

NINE

MAGIC GATES, EL NIÑO AND LA NIÑA

Global warming's effect on Earth's climate is a bit like a finger on a light switch. Nothing happens for a while but, if you increase the pressure, at a certain point a sudden change occurs, and conditions flick from one state to another.

Climatologist Julia Cole refers to the leaps made by climate as 'magic gates', and she argues that since temperatures began rising rapidly in the 1970s our planet has seen two such events—in 1976 and 1998.

The idea that Earth passed through a climatic magic gate in 1976 originated on the faraway coral atoll of Maiana in the Pacific nation of Kiribati. In fact it originated specifically in one of the oldest corals ever found—a 155-year-old *Porites*—that lived and grew there. When researchers drilled a section out of this coral they discovered a detailed record of climate change extending back to 1840.

The magic gate of 1976 could be seen in a sudden and sustained increase in sea surface temperature of 0.6°C, and a decline in the ocean's salinity (salt level) of 0.8 per cent.

Between 1945 and 1955 the temperature of the surface of the tropical Pacific commonly dipped below 19.2°C, but after the magic gate opened in 1976 it has rarely been below 25°C. 'The western tropical Pacific is the warmest area in the global ocean and is a great regulator of climate,' says Martin Hoerling, a climate researcher. It controls most tropical rainfall and the position of the Jet Stream, the powerful current of air high in the atmosphere whose winds bring snow and rain to North America.

In 1977 *National Geographic* ran a feature on the crazy weather of the previous year, which included unprecedented mild conditions in Alaska and blizzards in the lower forty-eight states of America. The immediate cause was a shift in the Jet Stream, but changes occurred as far afield as southern Australia and the Galápagos Islands which lie in the Pacific Ocean on the equator a thousand kilometres off the South American coast. The changes there affected evolution.

Charles Darwin visited the Galápagos Islands in the 1830s. He used its finches to illustrate his theory of evolution by natural selection. He could do this because the isolation of the islands had allowed its plants, birds and animals to develop under different circumstances. Since then the region has been a mecca for biologists, who established research stations to monitor its living creatures.

Scientists studying birds watched helplessly as the 1977 drought all but exterminated a species of native finch on one of the islands. Of the population of 1300 that existed before the drought, only 180 survived, and these were all individuals with the largest beaks, which enabled them to feed by cracking tough seeds.

Of those 180 survivors, 150 were males. When the rains finally came the male finches found themselves facing tough competition for mates. Again, it was those with the biggest beaks that won out. With this double whammy, a measurable shift in the beak size occurred on the island population. Since they had 150 years' worth of beak measurements to look back on, biologists felt they were witnessing the evolution of a new species.

The 1998 magic gate is tied up with the El Niño–La Niña cycle, a two- to eight-year-long cycle that brings extreme climatic events to much of the world.

The name El Niño, which in Spanish refers to the Christ child, was coined by Peruvian fishermen who noticed that a warm current often visited their fishing-grounds at Christmas. La Niña means the girl child and refers to a cooling period in the ocean off South America.

During the La Niña phase, winds blow westwards across the Pacific, pushing the warm surface water towards the coast of Australia and the islands lying to its north. With the warm waters shifted westwards, the cold Humboldt Current is able to surface off the Pacific coast of South America, carrying with it nutrients that feed the most prolific fishery in the world, the anchovetta.

The El Niño part of the cycle begins with a weakening of tropical winds, allowing the warm surface water to flow back eastwards, overwhelming the Humboldt and releasing humidity into the atmosphere which brings floods to the normally arid Peruvian deserts. Cooler water now upwells in the far western Pacific. It does not evaporate as readily as warm water, and so drought strikes Australia and southeast Asia.

When an El Niño is extreme enough, it can devastate two-thirds of the globe with droughts, floods and other extreme weather.

The 1997–98 El Niño year has been immortalised by the World Wide Fund for Nature (now the WWF) as 'the year the world caught fire'. Drought had a stranglehold on a large part of the planet. Fires burned on every continent, but it was in the normally wet rainforests of southeast Asia that they reached their peak. There over 10 million hectares burned, of which half was ancient rainforest. On the island of Borneo 5 million hectares were lost—an area almost the size of the Netherlands.

Many of the burned forests will never recover on a time scale meaningful to human beings. The impact on Borneo's unique fauna will probably never be fully known.

As greenhouse gases build up in the atmosphere we will experience persistent El Niño-like conditions.

Severe El Niño events can permanently alter the climate. The 1998 event released enough heat energy to spike the global temperature by around 0.3°C. Since then the waters of the central western Pacific have frequently reached 30°C, while the Jet Stream has shifted towards the North Pole. The new climatic regime also seems prone to generating more extreme El Niños.

Researchers wishing to document the response of nature to climate change often turn to the jottings of birdwatchers, fishermen and other nature watchers. Some of these records are very long—one English family recorded the date they heard the first frog and toad croaks on their estate every year between 1736 and 1947.

Prior to 1950 there is little evidence of any trend in these records, but over the last 55 years, right around the globe, a very strong pattern has emerged. Species have shifted towards the Poles by an average of around six kilometres per decade. They have retreated up mountain-sides at the rate of 6.1 metres per decade. And spring activity has advanced by 2.3 days per decade.

These trends accord with the scale and direction of temperature increases brought about by greenhouse gas emissions and have been hailed as a global 'fingerprint of climate change'. Such trends are so rapid and decisive it's as if the researchers had caught CO_2 in the act of driving nature Polewards with a lash.

Tiny marine organisms called copepods, for example, have been detected up to 1000 kilometres from their usual habitat. Thirty-five non-migratory species of Northern Hemisphere butterflies have flown northward, some by as much as 240 kilometres, while at the same time becoming extinct in the south. Even tropical species are on the move, with Costa Rica's lowland birds extending 18.9 kilometres northward over a twenty-year period.

With so many species relocating, it's inevitable that human changes to the environment will block their way.

Edith's checkerspot butterfly has a distinctive subspecies which inhabits northern Mexico and southern California. Increased temperatures in spring have caused the plant that its caterpillars feed on—a type of snapdragon—to wilt early, starving the larvae so they cannot pupate. The butterflies might have migrated to the north if the urban sprawl of San Diego didn't stand in their path. With only 20 per cent of their original habitat now able to support them, these butterflies may not be around next century.

The early onset of spring activity is a key clue to climate change. In the bird world the common murre, a Northern Hemisphere seabird, has begun to lay its eggs on average twenty-four days earlier each decade over the period its nesting has been studied. In Europe, numerous plant species have been budding and flowering 1.4 to 3.1 days earlier per decade, while their relatives in North America have been doing so 1.2 to two days earlier. European butterflies are appearing 2.8 to 3.2 days earlier per decade, while migrating birds are arriving in Europe 1.3 to 4.4 days earlier per decade.

As some species shift rapidly in response to climate change, others are left behind. A key food item might arrive too late to be of use to a predator, or move too far north for that predator to use it.

The caterpillars of the European winter moth eat only young oak leaves. But oaks and moths have different cues to tell them when spring has arrived. Warming weather causes the moth's eggs to hatch, but the oaks count the short cold days of winter as their guide for when to put out their leaves.

Spring is warmer than it was twenty-five years ago, but the number of cold days in winter hasn't changed. As a result, the winter moths now hatch up to three weeks before the oaks bear their first leaves. Because the caterpillars can survive only two to three days without food, there are now far fewer of them. Those that do survive generally grow faster because there is less competition for food, meaning the birds have less time to find them.

In this illustration, it seems likely that natural selection will act upon the moth to alter the timing of its hatching, but this will occur only through the mass deaths of early hatching caterpillars, and for several decades at least we can expect the species to be rare.

Will the birds, spiders and insects that eat the moths survive? If they can't it's another example of how all around the world climate change is tearing apart the delicate web of life.

Over the past few decades, breeding newts have been entering European ponds earlier, while frogs have not. This means that the newt tadpoles are well grown when those of the frogs hatch from their eggs. This allows the newts to eat large numbers of the frogs' young, which is having an impact on frog numbers.

Some reptiles face far more direct threats from global warming, for the sex of their young is determined by the temperature at which the eggs are incubated. For the North American painted turtle higher temperatures mean fewer males are born. If winter temperatures were to rise even slightly above their present high level, the creatures may find themselves with an all-female population.

A very different climate change impact was recently detected in Lake Tanganyika, Africa, one of the world's oldest and deepest freshwater bodies. Located just south of the equator, it's home to a host of unique species. Like most lakes, its waters are layered, with the warmest water on top. This can prevent the mixing of the oxygen-rich upper layers with the nutrient-rich ones below. Plants in the sunlit layers can be starved of nutrients, and those in the deeper layers can be starved of oxygen. In the past, this layering was seasonally broken down by the southeast monsoons, which stirred its waters and stimulated the spectacular variety of life it supports.

Since the mid-1970s, however, climate change has warmed its surface layers so much that the monsoons are no longer strong enough to mix the water. Inevitably, the plankton on which most lake life depends has now declined to less than one-third of its abundance of twenty-five years ago.

The spectacular spined snail, which is found only in the lake, has lost two-thirds of its habitat; today it lives only at depths of 100 metres or less, whereas twenty-five years ago it ventured three times as deep. These changes, scientists warn, threaten a collapse of the lake's entire ecosystem.

All over the world the surfaces of lakes are warming, preventing the mixing of their waters and threatening the basis of their productivity.

Even remote rainforest is being affected by global warming.

In areas of the Amazon far distant from any direct human influence, the proportion of trees that make up the canopy is changing. Spurred on by increased CO_2 levels, fast-growing species are powering ahead, crowding out slower-growing plants. This diminishes the rainforest's biodiversity, as the birds and animals that depend on the slower-growing species for food vanish, along with their resources.

One of the most important natural divisions on our planet is Wallace's Line. To the west lies Asia with its tigers and elephants, while to the east is a region, centred on Australia, known as Meganesia, which has an ancient and distinctive flora and fauna, including many marsupials.

The richest habitat in all of Meganesia is the mountainous oak forests of New Guinea. During the oak-fruiting season, the rich humus of the forest floor is littered with large, shiny brown acorns. If you pick one up, you'll most likely discover that it has been chewed, for these forests are home to more species of possum and giant rat than anywhere on Earth, and they love to snack on acorns.

Me with a baby giant woolly rat in the Nong River rainforests of central New Guinea in 1985. This creature's habitat no longer exists.

In 1985, when I first saw these wondrous forests—in the Nong River Valley north of Telefomin near the centre of the island—they stretched before me into the blue distance, an unbroken stand of wilderness. I was the first mammalogist ever to work in that area, a rare privilege. It was home to many unusual species, some of which were unique to the region and unknown to science.

One such creature was a greyish, cat-sized possum with large brown eyes, small paws and a short tail, which the Telefol people (who sometimes journeyed into the valley to hunt) knew as *matanim*. Talking to hunters, I gathered it had a singular diet of fig leaves, fruit and the rotten wood of certain trees.

The Nong isn't the easiest place in the world to reach, so in 2001 when I had the opportunity to return I jumped at it. You might imagine how excited I was, but even before the helicopter landed my spirits had plummeted. The entire valley, along with the surrounding peaks, had been transformed into a vast grove of vegetable tombstones.

Later, my old Telefol friends told me that during the last half of 1997 little or no rain fell, and the cloudless sky cast bitter frosts that killed the trees. By New Year, the remains of the forest had been baked to a crisp and its floor lay covered with dead leaves. When it came, the fire raced down through the valley and up onto the adjacent peaks. It burned for months, and even a year later it was likely to flare up from moss and dead plant matter buried deep underground.

These events had devastated the region, driving the wild animals from their haunts. The numbers of marsupial jaws kept by hunters as trophies showed that the environmental catastrophe had made the last untouched refuges accessible to humans. Strings of hundreds of jaws of the larger and rarer creatures such as tree-kangaroos, possums and giant rats hung from hearths, proof that even mediocre hunters had been assured of success.

Was there, hanging among those prizes, I wondered, the jawbones of the very last matanim on Earth?

It would take years of research to confirm the presence or absence of such a rare and elusive animal. But from what I saw on my visit in 2001, I think that its survival would have to be counted as a miracle.

TEN

PERIL AT THE POLES

In the final days of 2004, the cities of the world received some astonishing news: beginning at its northern tip, Antarctica was turning green.

Antarctic hair-grass usually survives as sparse tussocks crouched behind the north face of a boulder or some other sheltered spot. Over the southern summer of 2004, however, great green meadows began to appear in what was once the home of the blizzard. It's an emblem of the transformations occurring at the polar ends of our Earth. Yet changes on land are insignificant compared to those occurring at sea, for the sea ice is disappearing.

The subantarctic seas are some of the richest on Earth, despite an almost total absence of the nutrient iron. The presence of sea ice somehow compensates for this: the semi-frozen edge between salt water and the floating ice promotes remarkable growth of the microscopic plankton that is the base of the food chain.

Despite the months of winter darkness the plankton thrive under the ice, allowing the krill that feed on them to complete their seven-year life cycle. And wherever there is krill in abundance there are likely to be penguins, seals and great whales.

Ever since 1976 the krill have been in sharp decline, reducing at the rate of nearly 40 per cent per decade. As the krill numbers have decreased, those of another major grazing species—the jelly-like salps—have increased. Salps were previously confined to more northerly waters. They don't require a great density of plankton to thrive; they can survive on the meagre pickings in the ice-free parts of the Southern Ocean. But salps are so devoid of nutrients that none of the Antarctic's marine mammals or birds finds it worthwhile feeding on them.

The reduction in krill numbers seems to coincide closely with the warming of the ocean and the reduction of sea ice. There is little doubt that climate change is damaging the world's most productive ocean, as well as the largest creatures that exist and that feed there.

Imagine what it would mean for the beasts of the Serengeti in Africa if their grasslands had been reduced by 40 per cent each decade since 1976? Imagine what it would mean if your own living space were slashed by 40 per cent each decade?

The emperor penguin population is now half what it was thirty years ago, while the number of Adelie penguins has declined by 70 per cent.

Emperor penguins are already declining as a result of climate change. Small changes to their precarious environment may well drive them to extinction.

Southern right whales have only recently begun to return to Australian and New Zealand shores, but they will no longer come, because they need to fatten up on winter krill if they are to travel to their birthing grounds in warmer waters. The humpbacks that traverse the world's oceans will no longer be able to fill their huge bellies, nor will the seals and penguins that frolic in southern seas.

Instead we'll have a defrosting cryosphere (the term scientists use to describe the icy areas of the Earth) and an ocean full of jelly-like salps.

The Antarctic is a frozen continent surrounded by an immensely rich ocean. The Arctic, on the other hand, is a frozen ocean almost entirely surrounded by land. It's also home to 4 million people. Most of the Arctic's inhabitants live on the fringe, and it's there, in places such as southern Alaska, that winters are 2°C to 3°C warmer than they were thirty years ago.

Among the most visible impacts of climate change anywhere on Earth are those caused by the spruce bark beetle. Over the past fifteen years it has killed some 40 million trees in southern Alaska, more than any other insect in North America's recorded history. Two hard winters are usually enough to control beetle numbers, but a run of mild winters in recent years has allowed them to explode.

Collared lemmings are superbly adapted to life in the cryosphere, for they survive even on the hostile northern coast of Greenland. They're the only rodents whose coat turns white in winter, and whose claws grow into two-pronged shovels for tunnelling through snow. They are so abundant they can migrate en masse in search of food, though it's not true that they commit suicide by running off cliffs.

Scientists predict that, if global warming trends persist, forests will expand northwards to the edge of the Arctic Sea and destroy the vast plains and frozen subsoil of the tundra. Several hundred million birds migrate to these regions to breed. As the forest moves north the great flocks look set to lose more than 50 per cent of their nesting habitat this century alone.

For the collared lemming, the tundra and life itself are inseparable. Experts say it will be extinct before the year 2100. Perhaps all we will have then is a folk memory of the small, suicidal rodent.

But the tragedy will be that the lemmings didn't jump. They were pushed.

The caribou (or reindeer as the species is known in Eurasia) is vital to the Inuit, the Arctic's indigenous people. The Peary caribou is a small, pale subspecies found only in west Greenland and Canada's Arctic islands. A season in the Arctic with less snow but more rain can be devastating. Autumn rains now ice over the lichens that are the creature's winter food supply, causing many to starve. The number of Peary caribou dropped from 26,000 in 1961 to 1000 in 1997. In 1991 it was classified as endangered, which meant that it couldn't be hunted, and so it became irrelevant to the Inuit economy.

The Saami people of Finland have noted a similar icing of the reindeer's winter food supply. As climate change advances, it seems that the Arctic will no longer be a suitable habitat for caribou.

Can we imagine the North Pole without reindeer?

If anything symbolises the Arctic it is surely *nanuk*, the great white polar bear. He is a wanderer and a hunter, and a fair match for man in the white infinity of his polar world.

Every inch of the Arctic lies within his grasp: he has been sighted two kilometres up on the Greenland ice cap, and purposefully striding the ice within 150 kilometres of the true Pole itself. For polar bears, having sufficient food to live means lots of sea ice. And the sea ice is disappearing at the rate of 8 per cent per decade.

Polar bears, it's true, will catch lemmings or scavenge dead birds if the opportunity presents itself, but it's sea ice and *netsik*—the ringed seal that lives and breeds there—that are at the core of the *nanuk* economy.

Netsik is the most abundant mammal of the far north and at least 2.5 million of them swim in its berg-cooled seas. Yet at times climatic conditions prevent them from breeding. In 1974 too little snow fell over the Amundsen Gulf for the seals to construct their snow-covered dens on the sea ice. So they left, some travelling as far as Siberia.

And the polar bears? Those that had enough fat to migrate followed the seals, but many could not keep up and starved.

The harp seals live in the Gulf of St Lawrence. This seal population is genetically separate from the rest of the species. Like the ringed seals, they can raise no pups when there is little or no sea ice present—which happened to them in 1967, 1981, 2000, 2001 and 2002. The run of pupless years that opened this century is worrying. When a run of ice-free years exceeds the reproductive life of a female seal—perhaps a dozen years at most—the Gulf of St Lawrence population will become extinct.

The great white bears are already slowly starving as each winter becomes warmer than the one before. A long-term study of 1200 individuals living around Hudson Bay reveals that they are already 15 per cent skinnier on average than they were a few decades ago.

With each year, starving females give birth to fewer cubs. Some decades ago triplets were common; they are now unheard of. And back then around half the cubs were weaned and feeding themselves at eighteen months, while today it's less than one in twenty. In some areas increasing winter rain may collapse birthing dens, killing both the mother and cubs sleeping within. And the early break-up of the ice can separate denning and feeding areas; if the young cubs cannot swim the distances to find food, they will starve to death. In the spring of 2006, for the first time Inuit began to find drowned polar bears: the ice is now too far from shore.

In creating an Arctic with dwindling sea ice, we are creating a monotony of open water and dry land. Without ice, snow and *nanuk*, what will it mean to be Inuit—the people who named the great white bear, and who understand him like no other? When *nanuk* is fit and well-fed he will strip the blubber from a fat seal, leaving the rest to the arctic fox, the raven, and the gulls.

As the Arctic fills with hungry white bears, what will

become of these other creatures? The ivory gull has already declined by 90 per cent in Canada over the past twenty years. At that rate, it will not see out the century. The *nanuk* is already on the way to joining the list of endangered species.

It looks as if the loss of the *nanuk* may mark the beginning of the collapse of the entire Arctic ecosystem.

If nothing is done to limit greenhouse gas emissions, it seems certain that around 2050 a day will dawn when no summer ice will be seen in the Arctic—just a vast, dark, turbulent sea. But before the last ice melts the bears will have lost their den sites, feeding grounds and migration corridors.

Perhaps a group of elderly bears will linger on, each year becoming thinner than the last. Or perhaps a dreadful summer will arrive when the denning seals are nowhere to be found. A few bears might survive for a time on a diet of lemming, carrion and sea-caught seals, but they'll be so thin that they will not wake from winter's sleep. So fast are the changes that there are likely to be few or no polar bears in the wild by around 2030.

The changes we're witnessing at the Poles are of the runaway type. Unless we act quickly the realm of the polar bear, the narwhal and the walrus will be replaced by the cold, ice-free oceans of the north, and by the great temperate forests of the Taiga (the largest habitat on Earth, that stretches across Canada, Europe and Asia).

You might think that the encroaching forests, by taking in CO_2 as they grow, would help slow climate change. Scientists estimate that this will be offset by the loss of *albedo* or whiteness. A dark green forest absorbs far more sunlight, and thus captures far more heat, than does snow-covered tundra. The overall impact of foresting the world's northern regions will be to heat our planet even more swiftly.

Once this has happened, no matter what humanity does about its greenhouse gas emissions, it will be too late for a reversal. After existing for millions of years, the north Polar cryosphere will have vanished forever.

ELEVEN

2050: THE GREAT STUMPY REEF?

Of all the ocean's ecosystems, none is more diverse or beautiful in colour and form than a coral reef. And none, the climate experts and marine biologists tell us, is more endangered by climate change.

Are the world's coral reefs really on the brink of collapse?

It's a question that matters to humanity, for coral reefs yield around US$30 billion in income each year, mostly to people who have few other resources.

But financial loss may prove to be a small thing. The citizens of five nations live entirely on coral atolls, while fringing reefs are all that stand between the invading sea and tens of millions of other people. Destroy these fringing reefs, and for many Pacific nations you have done the equivalent of bulldozing Holland's dykes.

One of every four inhabitants of the oceans spends at least part of its life cycle in coral reefs. Such biodiversity is made possible by both the complex architecture of the corals, which provide many hiding places, and the lack of nutrients in the clear, tropical water.

Low levels of nutrients can promote great diversity. The best example of this is seen on the infertile sand plains of South Africa's Cape Province, where 8000 species of shrubby flowering plants co-exist in a mix as diverse as that of most rainforests.

The coral reefs are the marine equivalent of South Africa's sand plains. The arch-enemy of coral reefs are nutrients, and disturbances that break down the structure of the reefs. Then only a few weedy species—mostly marine algae—can proliferate.

When Alfred Russel Wallace sailed into Ambon Harbour in what is now eastern Indonesia in 1857, he saw:

> one of the most astonishing and beautiful sights I have ever beheld. The bottom was absolutely hidden by a continuous series of corals, sponges, actiniae, and other marine productions, of magnificent dimensions, varied forms, and brilliant colours. The depth varied from about twenty to fifty feet, and the bottom was very uneven, rocks and chasms, and little hills and valleys, offering a variety of stations for the growth of these animal forests. In and out among them moved numbers of blue and red and yellow fishes, spotted and banded and striped in the most striking manner, while great orange or rosy transparent medusae floated along near the surface. It was a sight to gaze

at for hours, and no description can do justice to its surpassing beauty and interest.

During the 1990s I often sailed down Ambon Harbour, yet saw no coral gardens, no medusae, no fishes, nor even the bottom. Instead, the stinking opaque water was thick with effluent and garbage. As I neared the town it just got worse, until I was greeted with rafts of faeces, plastic bags, and the intestines of butchered goats.

Ambon Harbour is just one among countless examples of coral reefs that have been devastated over the course of the twentieth century. Today, the practice of overfishing—including fishing with explosives and poisons—threatens reef survival. Disturbing reef biodiversity can also lead to outbreaks of plague species, such as the crown of thorns starfish. Another problem is the runoff of nutrients from land-based agriculture and polluted cities, which has helped degrade even protected places such as Australia's Great Barrier Reef.

During the 1997–98 El Niño, when the rainforests of Indonesia burnt like never before, the air was thick for months with a smog cloud rich in iron. Before those fires, the coral reefs of southwestern Sumatra were among the most diverse in the world, boasting more that 100 species of hard corals, including massive individuals over a century old. Then, late in 1997, a 'red tide' appeared off Sumatra's coast. The colour was the result of a bloom of minute organisms that fed on the iron in the smog. The toxins they produced caused so much damage it will take the reefs decades to recover, if indeed they ever do.

The smog cloud generated over southeast Asia during the 2002 El Niño was even larger—it was the size of the United States. On such a scale smog can cut sunlight by 10 per cent, and heat the lower atmosphere and ocean. Algal blooms are now devastating coastlines from Indonesia to South Korea and causing hundreds of millions of dollars worth of damage to aquaculture and corals alike. Recovery seems unlikely for any east Asian coral reef.

High temperatures lead to coral bleaching. To understand how we need to examine a reef far from human interference, where warm water alone is causing change. Myrmidon Reef lies far off the coast of Queensland, and the only people who go there are the scientists who survey it every three years. When they last went, in 2004, it looked 'as though it's been bombed'. This was the result of the reef crest being severely bleached, leaving a forest of dead, white coral. Only on the deeper slopes did life survive.

Coral bleaching occurs whenever sea temperatures exceed a certain threshold. Where the hot water pools the coral turns a deathly white. If the heating is temporary the coral may slowly recover, but when the heat persists it dies. Coral bleaching was little heard of before 1930, and it remained a small-scale phenomenon until the 1970s. It was the 1998 El Niño that triggered the global dying.

The Great Barrier Reef is the most vulnerable reef in the world to climate change. In all, 42 per cent of it was bleached in 1998, with 18 per cent suffering permanent damage.

In 2002, with the renewal of El Niño conditions, a pool of warm water around half a million square kilometres developed over the Great Barrier Reef. This triggered another massive bleaching event that on some inshore reefs killed 90 per cent of all reef-forming corals, and left 60 per cent of the Great Barrier Reef affected. In the few patches of cool water which remained, the coral was undamaged.

And 2006 looked liked it was going to be another dreadful year for the reef, but then Cyclone Larry arrived. It took enough heat from the ocean to stall the bleaching event, using the heat energy to power devastating winds that damaged or destroyed 50,000 homes in Queensland. It was a terrible price to pay to secure the reef, for at least another year.

A panel of seventeen of the world's leading coral reef researchers warned that by 2030 catastrophic damage will have been done to the world's reefs, and by 2050 even the most protected of reefs will be showing massive signs of damage.

According to reef scientists, a further rise of 1°C in global temperature will cause 82 per cent of the Great Barrier Reef to bleach and die; 2°C will bleach 97 per cent

Gobiodon species C. This small fish is a native of Papua New Guinea. The destruction of its reef habitat means it is now restricted to a single coral head.

of it; and after 3°C there will be 'total devastation'.

It takes the oceans around three decades to catch up with the heat accumulated in the atmosphere, so it may be that four-fifths of the Great Barrier Reef is one vast zone of the living dead—just waiting for time and warm water to catch up with it.

Extinctions caused by climate change are almost certainly under way on the world's reefs, and a tiny species of coral reef-dwelling fish known as *Gobiodon* species C may be symbolic of them. Most of the habitat used by this tiny creature was destroyed by bleaching during the 1997–98 El Niño, and it can now be seen only on one patch of coral in one lagoon in Papua New Guinea.

'Species C' indicates that it has not yet been formally named, and it may become extinct before this can happen. It isn't an exaggeration to say that we need to multiply the loss of this one little fish a thousandfold to gain a sense of the cascade of extinctions that is occurring right now.

A survey conducted in 2003 revealed that live coral cover had dropped to less than 10 per cent on half of the area of the Great Barrier Reef. Significant damage was evident in even the healthiest sections. Public outrage made political action inevitable, and the Australian government announced that 30 per cent of the reef would be protected. This meant that commercial fishing would be banned, and other human activities severely curtailed, in the newly protected zone.

But it is not fishing or tourists that is killing the Great Barrier Reef. It is spiralling CO_2 emissions. And Australians produce more CO_2 per person than the people of any other nation on Earth.

If we are to have a chance of saving these wonders of the natural world we need to reduce our greenhouse gas emissions now.

TWELVE

A WARNING FROM THE GOLDEN TOAD

In our story we are yet to meet a single species that has definitely become extinct because of climate change. In the regions where it is likely to have occurred, such as New Guinea's forests and coral reefs, there's been no biologist on hand to document the event. In contrast, there are many researchers at the Monteverde Cloud Forest Preserve in Costa Rica, Central America, where the Golden Toad Laboratory for Conservation is located.

Soon after our fragile planet passed through the climatic magic gate of 1976, abrupt and strange events were observed by the ecologists who spend their lives working in these pristine forests.

During the winter dry season of 1987, the frogs that live in the mossy rainforests one and a half kilometres above the sea began to disappear. Thirty of the fifty species known to inhabit the 30-square-kilometre study site vanished. Among them was a spectacular toad the colour of spun gold. The golden toad lived only on the upper slopes of the mountain.

At certain times of the year crowds of the brilliant males gathered around puddles on the forest floor to mate.

The golden toad was discovered and named in 1966, although the Indians knew about it long before. They have myths about a mysterious golden frog that is very difficult to find, but should anyone search the mountains for long enough to find one they will obtain great happiness. Their stories tell of one man who found the frog but let it go because he found happiness too painful to bear. Another released the creature because he didn't recognise happiness when he had it.

Only the males are golden; the females are mottled black, yellow and scarlet. For much of the year it's a secretive creature, spending its time in burrows amid the mossy roots of the woodland. Then, as the dry season gives way to the wet in April–May, it appears above ground en masse, for just a few days or weeks. With such a short time to reproduce, the males fight with each other for top spot and take every opportunity to mate—even if it's only with a field worker's boot.

In her book *In Search of the Golden Frog*, amphibian expert Marty Crump tells us what it was like to see the creature in its mating frenzy:

> I trudge uphill...through cloud forest, then through gnarled elfin forest...At the next bend I see one of the most incredible sights I've ever seen. There,

congregated around several small pools at the bases of dwarfed, windswept trees, are over one hundred Day-Glo golden orange toads poised like statues, dazzling jewels against the dark brown mud.

On 15 April 1987 Crump made a note in her field diary that was to have historic significance:

> We see a large orange blob with legs flailing in all directions: a writhing mass of toad flesh. Closer examination reveals three males, each struggling to gain access to the female in the middle. Forty-two brilliant orange splotches poised around the pool are unmated males, alert to any movement and ready to pounce. Another fifty-seven unmated males are scattered nearby. In total we find 133 toads in the neighbourhood of this kitchen sink-sized pool.

On 20 April:

> Breeding seems to be over. I found the last female four days ago, and gradually the males have returned to their underground retreats. Every day the ground is drier and the pools contain less water. Today's observations are discouraging. Most of the pools have dried completely, leaving behind desiccated eggs already covered in mold. Unfortunately, the dry

weather conditions of El Niño are still affecting this part of Costa Rica.

As if they knew the fate of their eggs, the toads attempted to breed again in May. This was, as far as the world knows, the last great toad orgy ever to occur. Despite the fact that 43,500 eggs were deposited in the ten pools Crump studied, only twenty-nine tadpoles survived for longer than a week, because the pools once again quickly dried.

The following year Crump was back at Monteverde for the breeding season, but this time things were different. After a long search, on 21 May she located a single male. By June, and still searching, Crump was worried: 'the forest seems sterile and depressing without the bright orange splashes of colour...I don't understand what's happening. Why haven't we found a few hopeful males, checking out the pools in anticipation?'

A year was to pass before, on 15 May 1989, a solitary male was again sighted. As it was sitting just three metres from where Crump made her sighting twelve months earlier, it was almost certainly the same toad.

For the second year running he held a lonely vigil, waiting for the arrival of his fellows. He was, as far as we know, the last of his species. The golden toad has not been seen since.

Other species at Monteverde were also affected. Two species of lizard vanished entirely. Today, the mountain's rainforests continue to be stripped of their jewels, with many reptiles, frogs and other fauna becoming rarer by the year. While still verdant enough to justify its name, the Monteverde Cloud Forest Preserve is beginning to resemble a crown that has lost its brightest gems.

Researchers began to study the records of temperature and rainfall. Eventually, in 1999, they announced that they had solved the mystery of the disappearance of the golden toad.

Ever since Earth passed through its first climatic magic gate in 1976, there were more and more mistless days each dry season on Monteverde, until they had joined into runs of mistless days. By the dry season of 1987, the number of consecutive mistless days had passed some critical threshold. Mist, you see, brings vital moisture. Its absence caused catastrophic changes.

Why, the researchers wanted to know, had the mist left Monteverde? Beginning in 1976 the bottom of the cloud mass had risen until it was above the level of the forest. The change had been driven by the abrupt rise in sea surface temperatures in the central western Pacific. A hot ocean had heated the air, elevating the condensation point for moisture. By 1987 the rising cloud-line was, on many days, above the mossy forest altogether, bringing shade but no mist. The golden toad has porous skin, and it likes to wander in daylight hours. It was exquisitely vulnerable to the new drier climate.

It's always devastating when you witness the extinction of a species. You are seeing the dismantling of ecosystems and irreparable genetic loss. It takes hundreds of thousands of years for such species to evolve.

The golden toad is the first documented victim of global warming. We killed it with our reckless use of coal-fired electricity and our huge cars, just as surely as if we had flattened its forest with bulldozers.

Since 1976 many researchers have observed amphibian species vanish before their eyes without being able to determine the cause. New studies indicate that climate change is responsible for these disappearances too.

In the late 1970s, a remarkable creature known as the gastric brooding frog disappeared from the mossy forests of southeastern Queensland. When it was first discovered in 1973 this brown, medium-sized frog astonished a researcher who looked into a female's open mouth—to observe a miniature frog sitting on her tongue! Not just the frog—scientists around the world were open-mouthed too.

The species is not a cannibal. It has bizarre breeding habits. The female swallows her fertilised eggs, and the tadpoles develop in her stomach until they metamorphose into frogs, which she then regurgitates into the world.

When this novel method of reproduction was announced,

Australia's gastric brooding frog nurtured its tadpoles in its stomach, which it somehow transformed from an organ of digestion into a brood chamber. The species may well be Australia's first victim of climate change.

some medical researchers understandably got excited. How did the frog transform its stomach from an acid-filled digesting device into a nursery? The answer might help doctors treat a variety of stomach complaints. Alas, they were unable to carry out many experiments, for in 1979—six years after humans learned of its existence—the gastric brooding frog vanished, and with it went another inhabitant of the same streams, the day frog. Neither has been seen since.

In the early 1990s, frogs began to disappear en masse from the rainforests of northern Queensland. Today some sixteen frog species (13 per cent of Australia's total amphibian fauna), have experienced falls in numbers. The decreases in rainfall experienced in eastern Australia over the past few decades cannot have been good for frogs. At least in the case of the gastric brooder and day frog, climate change is the most likely cause of their disappearance.

Now almost a third of the world's 6000-odd species of amphibians is threatened with extinction. Some scientists believe shallower breeding ponds—due to El Niño-like conditions—may be to blame. Fungal diseases are also contributing to the extinctions, and climate change is altering conditions in such a way that the fungus is flourishing.

Climate change seems to be the hidden cause of this wave of amphibian extinction.

THIRTEEN

RAINFALL

From the Poles to the equator our Earth spans a range of temperatures from around 40°C below zero to 40°C above. Air at 40°C can hold 470 times as much water vapour as air at –40°C. It's this fact that condemns our Poles to be great frozen deserts. And it's why, for every degree of warming we create, our world will experience an average 1 per cent increase in rainfall.

This extra rain is not evenly distributed. Instead, rain is appearing at unusual times in some places, and disappearing in others.

Over large parts of the world, rainfall is increasing. But more rain is not necessarily a good thing. As our planet warms, more rain will fall at high latitudes in winter, turning snow to ice and mush, which is bad news for the inhabitants of the Arctic. Further south, increasing winter rain is also bringing unwelcome change: in 2003 it triggered a deadly avalanche season in Canada, while the British spring of 2004 was so wet that in many regions hay-making was difficult or impossible.

Climate change will tip some regions into perpetual rainfall shortages.

Some may be transformed into new Saharas, or at least into regions unfit for human habitation. A lack of rainfall is often referred to as a 'drought', yet droughts don't last forever. In the areas we're going to explore the rain will probably never return. Instead, what has occurred is a rapid shift to a new, drier climate.

The first evidence of this emerged in Africa's Sahel region during the 1960s. The area affected was huge—an enormous tract of sub-Saharan Africa extending from the Atlantic Ocean to Sudan. It takes in a number of countries including Senegal, Nigeria and Ethiopia, Eritrea and Somalia. Four decades have now passed since the sudden decline in rainfall, and there is little sign that the life-giving monsoon rains will return.

Even before the decline, the Sahel was a region of marginal rainfall where life was tough. In areas with better soils and more rain, farmers made a living out of their fields. In the drier wastes, camel herders followed their semi-nomadic round in pursuit of feed for their animals.

The decreased rainfall has made life difficult for both groups: herders struggle to find grass in what is now a true desert, while the farmers rarely get sufficient rain to stir their fields to life. The world's media periodically shows images of the result—starving camels and desperate families struggling in a dust-filled wasteland.

I remember as a child seeing these images on television, and hearing about how an escalating population had caused this human misery. For decades the West has reassured itself that this disaster was brought on by the African people themselves. The argument was that overgrazing by camels, goats and cattle, as well as people gathering firewood, had destroyed the region's thin covering of vegetation, exposing its dark soil and changing its albedo. As this manmade 'drought' lengthened, the soil began to blow away. It's a view of things shared by many environmentalists and aid workers; but it is wrong in almost every respect.

The true origin of the Sahel disaster was revealed when climatologists in the US published a painstaking study that used computer models to simulate rainfall in the region between 1930 to 2000.

The model revealed that the amount of human-caused land degradation was too small to have triggered the dramatic climate shift. Instead, a single change was responsible for much of the rainfall decline: rising sea-surface temperatures in the Indian Ocean, caused by increased greenhouse gases.

The Indian Ocean is the most rapidly warming ocean on Earth. As it warms, the conditions that generate the Sahelian monsoon weaken. That's why the Sahel lost much of its rainfall.

There is growing evidence that the Sahel climate shift will eventually affect the entire world. Around half of the global dust in the air today originates in arid Africa, and the impact of the drying is so great that the planet's atmospheric dust loading has increased by a third. Dust is important stuff, because its tiny particles can scatter and absorb light, thereby lowering temperature. These particles also carry nutrients into the ocean and to distant lands, assisting the growth of plants and plankton, and thereby increasing the absorption of CO_2. The precise impact of this additional dust on world climate is uncertain, but it's likely to be substantial.

The citizens of the industrialised world tend to feel that their technology will protect them from Sahelian-scale disasters, but nature has been busy proving them wrong.

Australia is a dry country, and Australians are obsessed with rainfall. The southwestern corner of Western Australia once enjoyed reliable rainfall. Traditionally the rain fell during the winter, with over 100 centimetres falling annually at some locations. This made the area famous for its farming. The western wheatbelt was one of the largest and most reliable centres of grain production in the entire continent. More recently vineyards have spread throughout the wetter areas, and they produce some of the finest wines made in the Southern Hemisphere.

Before European settlement, most of the southwest was blanketed in a tough, spiny heath-like vegetation known as kwongan. Following the winter rains the kwongan region was transformed into a vast wildflower garden. Only in the tropical rainforest and a similar region in South Africa are more species jammed into a single hectare.

During the first 149 years of European settlement in the southwest (1829–1975) the reliable winter rainfall brought prosperity and opportunity. People cleared the kwongan for farmland. But from 1976 things changed and ever since the region has endured a decrease in rainfall averaging 15 per cent. Climate models indicate that about half the decline results from global warming, which has pushed the temperate weather zone southward.

Researchers believe the other half may result from destruction of the ozone layer, which has cooled the stratosphere over the Antarctic. This has hastened the circulation of cold air around the Pole and drawn the southern rainfall zone even further southwards.

The loss of rain was felt immediately on farms, particularly on the region's margins where variation of a few tens of millimetres makes the difference between a good harvest and failure. In these areas wheat is the principal crop, and it's grown in an unusual manner. In the 1960s the goal of the western farmers was to clear a million acres of kwongan scrub a year. When the bulldozers had done their work farmers found themselves staring at sterile stretches of sand—some of the most infertile soil to be found anywhere on Earth—for here, as in rainforests, the region's natural wealth was bound up in its vegetation.

But this was what the farmers wanted. Wheat-growing in the southwest was a gigantic version of hydroponic gardening: farmers drilled in their wheat, dusted the sterile sand with nutrients, then waited for the never-failing winter rains to add water.

By 2004, after decades of nature refusing to add water, the wheat-growing region began to withdraw westwards, replacing dairying in country once considered too wet for the crop. As conditions worsen over the coming century, the Indian Ocean will become the ultimate barrier to this process: one high-rainfall activity after another will face being pushed into the sea.

That 15 per cent reduction in rainfall hides an even greater catastrophe: winter rainfall has in reality declined by more than that, while summer rainfall (which is far more erratic) has increased. Because summer rains cannot be depended upon, farmers do not plant summer crops, so the rain falls on bare fields, allowing the water to soak down to the watertable. There it meets salt, which steady westerly winds have been blowing in from the Indian Ocean for millions of years.

Under every square metre of this land lies an average of between 70 and 120 kilograms of salt. Before land-clearing this didn't matter, for the diverse native vegetation of the kwongan used every drop of water that fell from the heavens, and the salt stayed in its crystalline form.

As the summer rains began to fall on the vacant wheat-fields, however, water far saltier than the sea began to creep upward, killing everything it touched. The first sign of trouble was a salty taste in the previously sweet creeks of the region. In many cases they quickly became undrinkable, their streamside vegetation died and within a decade or two they had turned into collapsed, salty drains.

Today, impoverished and bankrupt farmers in the west of Australia are facing the worst case of dry-land salinity in the world. Neither science nor government has been able to provide solutions, and the damage bill is in the billions.

The government itself admits that the area of salt-affected land in Western Australia is increasing at a rate of one football field per hour. Roads, railways, houses and airfields are now besieged by salt. Unless the original vegetation can be returned and induced to grow in the drier and saltier conditions that now exist, there appears to be no hope of a turnaround.

Western Australia's capital is Perth, a thirsty city of 1.5 million people and the world's most isolated metropolis. There, a taxi driver might be a bankrupt wheat farmer scraping together a living as he tries to sell a now useless farm. The decline in winter rainfall also means less water in Perth's catchments. Since 1975 the rain has tended to fall in light showers that soak into the soil and do not reach the dams.

Over most of the twentieth century an average of 338 gigalitres of water per year had flowed into the dams that quench the city's thirst. (A gigalitre is a billion litres, the equivalent of 500 Olympic-sized swimming pools.) But between 1975 and 1996 the average was only 177 gigalitres—representing a cut of around 50 per cent to the city's surface water supply. Between 1997 and 2004 it fell to just 120 gigalitres—little more than a third of the flow received three decades earlier.

Severe water restrictions were put in place in 1976, but the situation was eased by drawing on a reserve of groundwater known as the Gnangara Mound. For a quarter of a century the city mined this subterranean water, but the failing rains meant that it was not being recharged. In 2001 Perth's dams received virtually no water, and by 2004 the situation of the Gnangara Mound was critical. The state's Environmental Protection Authority warned that extracting more water from it would threaten some species with extinction. Today, the western swamp tortoise, which is a living fossil, only survives because water is pumped into half of its habitat.

By early 2005, nearly thirty years after the crisis first emerged, Perth's water experts predicted a one in five chance of a 'catastrophic failure of supply'—which means no water coming out of the tap. In that case the city would have no choice but to squeeze what water it could out of the Gnangara Mound, destroying much ancient and wondrous

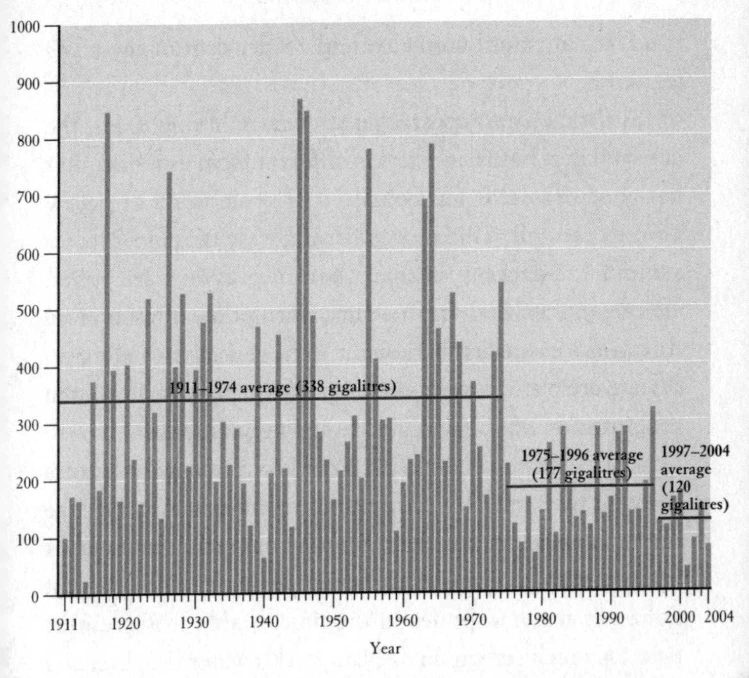

This shows the water flow into Perth's catchments between 1911 and 2004. Large reductions followed the magic gate years 1976 and 1998, and the city has lost two-thirds of its surface water supply over the last thirty years.

biodiversity, and *still* the fix would only be temporary.

Plans have now been laid for a desalination plant at a cost of around A$350 million, making it one of the largest in the Southern Hemisphere. The plant will suck water from the ocean and remove the salt. This takes so much power that the water is sometimes referred to as liquid electricity, so it's good that the plant will be powered by wind. It will supply only around 15 per cent of the city's water.

Australia's east coast is no stranger to drought, but the dry spell that began in 1998 is different from anything that has gone before. It has consisted of seven years of below average rainfall. This is a hot drought, with temperatures around 1.7 degrees warmer than in previous droughts, making it of exceptional hostility. The decline of rainfall on Australia's east coast is thought to be caused by a climate-change double whammy—the loss of winter rainfall and the prolongation of El Niño-like drought conditions.

Cities such as Sydney lack the groundwater resources enjoyed by Perth. Its only buffer against below-average rainfall is its dams, so that a decline in stream flow means water hardship. Sydney's water supply is one of the largest domestic water supplies in the world, able to store four times as much per capita as New York's water supplies, and nine times as much as London's.

Yet even this storage has proved insufficient. Between 1990 and 1996 the total flow into all eleven of Sydney's dams had averaged 72 gigalitres per month. But by 2003 this had dropped to just 40 gigalitres, a decline of 44 per cent. The situation remains critical. Sydney's 4 million residents have around two years' supply of water in storage.

There is hardly a city in Australia that is not facing a water crisis of some sort.

Across the Pacific Ocean, parts of the American west are in their sixth year of drought. Such dry conditions have not been seen in the region for around 700 years, when the American southwest was even warmer than it is today. This suggests a relationship between drought and warmer conditions. As with the Sahel, the link seems to lie in rising ocean temperatures.

The drought conditions in the American west are frequently portrayed in the media as being part of a natural cycle. But the changes are the same as those expected to result from global warming, and as those observed during warm times in the past. Climate change has the potential to cause drought almost anywhere on the planet.

Much of the water in the American southwest comes in the form of winter snow that accumulates on its high mountains. Because it melts over the spring and summer, it provides stream flow when most needed by farmers. The snowpack has offered an inexpensive form of water storage that has minimised the need for dams. The amount of snow that falls has always varied from year to year, but over the last fifty years there has been a decline in the average amount received.

If this trend continues for another five decades, western snowpacks will reduce by up to 60 per cent in some regions, which could cut summertime stream flow in half. This will devastate not just water supplies, but hydropower and fish habitats as well.

Over the past fifty years, the southwest region has warmed by 0.8°C—slightly more than the global average. This has reduced the snowpack because the higher temperatures are melting it before it can consolidate. On the whole the snowpack is melting earlier, which means that the peak of runoff into streams is now occurring three weeks sooner than in 1948.

This leaves less water for the height of summer, when it's most needed, but increases water flow in winter and spring, which may lead to more flooding. Temperatures in the region are set to rise between 2°C and 7°C over this century unless we significantly reduce CO_2 emissions. If we do nothing, most streams will eventually flow in winter, when the water is least needed.

'So what?' many people may say. 'We'll just build more dams.'

It is possible that, as the crisis deepens, this is what they will do. But there are a limited number of sites suitable for dams in the region, and dams mean that farmers will pay for water storage that was once provided by nature. Besides, the changes under way are so vast that even a new program of dam-building is insufficient to counter them. Researchers forecast that snowpack changes could lower farm values by 15 per cent, costing billions.

The biggest problem, however, is surely to do with the cities of the US west, which will suffer dwindling water supplies. These vast metropolises are impossible to relocate and some—as it was with the ancient cities of Mesopotamia—may have to be abandoned if the rate of change accelerates. This may sound extreme, but we are only at the beginning of the West's water crisis.

Five thousand years ago, when the American southwest was a little warmer and drier even than it is today, the Indian cultures that had flourished across the region all but vanished. Only when conditions cooled again was the region habitable. For more than a millennium the southwest was like a big ghost town.

FOURTEEN

EXTREME WEATHER

In 2003 climate scientists announced that, over just a few years, the tropopause (the boundary in the atmosphere between the troposphere and the stratosphere) had risen by several hundred metres. The cause was both warming and expansion of the troposphere due to extra greenhouse gases, as well as cooling and contraction of the stratosphere due to ozone depletion.

Why should this small adjustment eleven kilometres above our heads worry us?

For the very good reason that climatologists now realise that the tropopause is where much of our weather is generated. Change it, and you alter not only weather patterns but extreme weather events as well.

There is no doubt that extreme weather events are becoming more frequent.

Here is just a small sample from the past few years: the most powerful El Niño ever recorded (1997–98), the most fatal hurricane in 200 years (Mitch, 1998), the hottest and deadliest European summer on record (2003), the first

South Atlantic hurricane ever (2002), unprecedented flooding in Mumbai, India (2005), the worst storm season ever experienced in the USA, the most economically devastating hurricane on record (Katrina, 2005), and Monica, the most powerful cyclone ever recorded in Australia (2006).

How could global warming make hurricanes and cyclones more powerful? The answer lies in the warming of the oceans, and the capacity of warmer air to hold water vapour, the fuel that drives these extreme storms.

Think of how on a hot day you perspire, and as your sweat evaporates it carries heat from your body into the air. It's a highly effective form of heat transfer, for the evaporation of just one gram of water from your skin is enough to transfer 580 calories of heat.

Think of the difference in scale of your body and an entire ocean, and you can sense the incredible power that heat energy derived from evaporation carries into the atmosphere.

So warm water can warm the air. And air warmed by climate change can carry so much extra heat. For every 10°C increase in its temperature, the amount of water vapour that the air can hold doubles; so at 30°C air can hold four times as much 'hurricane fuel' as air at 10°C.

Perhaps the most marked change in hurricanes since around 1950—when global warming began to be felt—is the change in their tracks. One of the best-documented examples of this comes from eastern Asia. The frequency of typhoons ravaging the East China and Philippine seas has decreased since 1976, but the number in the South China Sea has increased.

Further westward, in the Arabian Sea and in the Bay of Bengal, there have been fewer typhoons, which is good news for the millions living near sea level in these regions. Another marked change has been noted at high latitudes in the Southern Hemisphere, where there has been a dramatic decrease in the number of cyclones occurring over the sub-Antarctic Ocean south of latitude 40, but a modest increase in the Antarctic Ocean.

There are disturbing signs that hurricanes are becoming more frequent in North America. In 1996, 1997 and 1999 the United States endured more than twice the number of hurricanes experienced annually during the twentieth century. What 1998's hurricanes lacked in numbers, they more than made up for in intensity.

Hurricane Mitch tore through the Caribbean in October 1998, killing 10,000 people and making 3 million homeless. With wind speeds of up to 290 kilometres per hour, Mitch was the fourth strongest Atlantic Basin hurricane ever recorded. It was also the deadliest storm to hit the Americas in 200 years; only the Great Hurricane of 1780 which killed at least 22,000 people was more severe in its impact.

The storms returned with a vengeance in 2004, when four major tropical storms crossed the Florida coast in quick succession, devastating large parts of the state. Many of the homes damaged by these storms are still uninhabitable.

In September 2005 Hurricane Katrina burst upon New Orleans and changed climate history. Then Rita shook Texas, and people at last began to wonder if these giant engines of destruction were generated by climate change.

As all hurricanes do, Katrina started as a thunderstorm, in this case in warm waters off the Bahamas. Then it became a tropical storm, a group of thunderstorms that circled until they developed a vortex.

Tropical storms intensify into hurricanes only where the surface temperature of the ocean is around 26°C or greater. This is because hot sea water evaporates readily, providing the volume of fuel—water vapour—required to power a hurricane.

Hurricanes are classified on the Saffir-Simpson Hurricane Scale, which runs from 1 to 5. Category 1 hurricanes lack the puff to do real damage to most buildings, but they can generate a 1.5-metre storm surge, flood coastlines and damage poorly constructed buildings.

Category 3 hurricanes are more dangerous. They generate winds of between 180 and 210 kilometres per hour, and can destroy mobile homes and strip trees of their leaves.

Category 5 hurricanes are an entirely different matter. When they make landfall, 250 kilometres per hour winds ensure that there are no trees or shrubs left standing. Nor are there a lot of buildings left. Storm surges exceeding

5.5 metres arrive around four hours before the eye of the storm hits, so flooding is far more extensive, and routes are obstructed, preventing people from fleeing.

When Katrina slammed into Florida on 25 August, it was a category 1 storm with wind speeds of 120 kilometres per hour. Katrina killed eleven people in Florida. Hurricanes often die out when they cross onto land, but somehow Katrina survived the transit of the Florida Peninsula and on 27 August emerged into the Gulf of Mexico.

During the summer of 2005 the surface waters of the northern Gulf were exceptionally hot—around 30°C. This is too warm to enjoy a swim. Large bodies of sea water don't get much hotter, and the Gulf waters are deep, which provides a large heat reservoir. Such waters yield vast volumes of water vapour. During its four-day passage through Gulf waters Katrina grew and grew, until it reached category 5.

By the time Katrina neared New Orleans it had subsided into a category 3 storm, the eye of which passed fifty kilometres east of the city. Thus, Katrina was not the fiercest of storms when it struck, nor did it score a direct hit on the city. Yet its impact was catastrophic.

Half a million people inhabited the city, large parts of which lay several metres below sea level—a key factor in its vulnerability. The levees that kept the waters of the Mississippi and Lake Pontchartrain at bay had been built with a kinder climate in mind, and could not withstand the

Hurricane Katrina, Category 5, as it approaches America's coastline. It had a devastating impact on New Orleans. A hurricane like this can generate waves up to 15 metres high, storm surges of 6 metres, and 2 billion tonnes of rain per day.

impact of the hurricane. With the number of very powerful hurricanes increasing over the past decade it was widely understood that the devastation of the city was only a matter of time. In October 2004 *National Geographic* ran a story outlining the dangers, and in September 2005 *Time* again listed what they were.

So many things went wrong in New Orleans. Poverty, high levels of gun ownership, and official corruption and incompetence, all combined to hinder the relief effort. Then there was the industrial pollution released by the storm surge and high winds. In a region that supplies and refines a considerable proportion of America's oil, spills were inevitable. Katrina flooded many of the 140 large petrochemical works that comprise Louisiana's 'cancer corridor'. This damage, of course, was magnified by Rita, which hit at the heart of the US petrochemical industry in Texas.

All of this teaches us that many of the most devastating impacts of any individual hurricane are not related to global warming. Whether Katrina was a little weaker or stronger, whether it struck fifty or 150 kilometres from the city, and whether it struck a week earlier or later, are all matters of chance.

But there is ample evidence that global warming is changing the conditions in the atmosphere and oceans in ways that will make hurricanes even more destructive in future.

Scientists have found that the total amount of energy released by hurricanes worldwide has increased by 60 per cent in the last two decades, and that more energy is going into the most powerful hurricanes.

Since 1974 the number of Category 4 and 5 hurricanes has almost doubled. Hurricanes are lasting longer, and hurricane seasons are extending. Today we have no hurricanes in winter because the sea is too cold. In a warming world this need not be so.

Hurricanes and cyclones focus attention on climate change in a way that few other natural phenomena do. And they have the potential to cause much more damage and kill many more people than the largest terrorist attack. Living with a heightened risk of such devastation should act as a constant reminder to us that the failure to combat climate change carries a high price indeed.

In the wake of hurricanes come floods. Since warmer air is able to hold more water vapour, the incidence of severe floods is rising and expected to rise further. In the summer of 2002 two-fifths of South Korea's annual rainfall fell in a week, causing such destruction that the nation had to mobilise its troops to help flood victims. At the same time China suffered floods of historic magnitude, with 100 million people affected.

The global increase in flood damage over recent decades has been profound. In the 1960s around 7 million people were affected by flooding annually. Today that figure stands at 150 million. And floods bring disease. Cholera breeds in the stagnant and polluted water, as do mosquitoes that can spread malaria, yellow fever, dengue fever and encephalitis. Even plague can benefit from the disturbance as fleas, rats and humans are brought together on higher ground.

Scientists have also found that the United Kingdom has experienced a significant increase in severe winter storms, a trend they predict will continue. This is linked to a warming climate: the 1990s were the warmest decade in central England since records began in the 1660s. The growing season for plants has been extended by a month, heatwaves have become more frequent and winters much wetter, with heavier rain.

On the continent more alarming events have occurred. The European summer of 2003 was so hot that, statistically speaking, such an extreme event should occur once every 46,000 years. During June and July of that year 30,000 people died when temperatures exceeded 40°C across much of Europe. Heatwaves kill a large number of people worldwide each year; even in the climatically turbulent US, heat-related deaths exceed those from all other weather-related causes combined.

The United States already has the most varied weather of any country on Earth, with more damaging tornadoes, flash floods, intense thunderstorms, hurricanes and blizzards than anywhere else. With the intensity of such events projected to increase as our planet warms, the people of the United States may have more to lose from climate change than any other large nation.

As we have seen from its abrupt decreases in rainfall, Australia too is suffering the effects of climate change: severe storms, an increase in the number of very hot days, an increase in night-time temperatures, a decrease in very cold days and a decrease in the incidence of frosts.

Some regions, such as around Alice Springs in central

Australia, have experienced an increase in temperature of more than 3°C over the twentieth century. There has also been an increase in the occurrence of intense cyclones, as well as severe low-pressure systems in southeastern Australia. The frequency of floods has also increased, particularly since the 1960s.

It's difficult to find two nations that have been more severely disadvantaged by climate change than the US and Australia.

Some regions of the world, in contrast, have so far recorded less change. In India, apart from Gujarat and western Orissa, there is less drought than twenty-five years ago. Only northwestern India is experiencing a marked increase in extremely hot days: the heatwaves there are causing many deaths. And in 2005 there were record monsoon rains and storms in Mumbai and surrounding regions, which brought devastating floods and damaged the offshore Mumbai high gas field.

One impact of global warming is being felt on all continents equally: all of them are shrinking. This is because, courtesy of heat and melting ice, the oceans are expanding. Is this a threat to humanity? How far will the waters rise, and at what speed?

FIFTEEN

RISING WATERS

Our species most likely began in the lake region of the African Rift Valley, where our ancestors foraged on the bounty of fish, shellfish, birds and mammals. We have sought to live close to water ever since, for water draws living things from near and far. Camp near a waterhole and sooner or later animals will come to drink. Almost instinctively, human beings have always preferred to live with a water view, especially if it includes a beach, a lake or a lawn cropped short as if by great grazing beasts. Real estate agents understand our housing preferences and the amount we are willing to pay for them.

Two out of every three people on Earth live within eighty kilometres of the coast. Yet in our subconscious we understand that the waters can rise over the land, making all of our hard-won real estate count for nothing.

Fifteen thousand years ago the oceans were at least 100 metres lower than they are today. North America was an empire of ice, exceeding even the Antarctic in the volume of frozen water it supported. As the great American ice caps melted they alone released enough water to raise global sea levels by seventy-four metres.

The sea rose rapidly until around 8000 years ago, when it reached its present level and conditions stabilised. All around the world people watched the waters rise, at times so fast as to change the coastline from year to year. Today, even a modest sea level rise would be disastrous, everywhere from Manhattan to the Bay of Bengal, for the human population along coastlines is dense.

Although it is not related to climate change, the catastrophic Asian tsunami of 2004 gives some indication of how devastating rising seas and turbulent weather might be. The Netherlands is already planning for the construction of a super-dyke to save it from the encroaching ocean. The Thames Barrier that protects London from disastrous tidal surges is to be strengthened. But countless millions of others live beside the sea—some on expensive estates, others in humble villages—and have no protection. In Bangladesh alone, more than 10 million people live within one metre of sea level. The last time the world was as warm as it's projected to be by 2050, sea levels were four metres higher than today.

All that remains of the great Northern Hemisphere ice caps today is the Greenland ice sheet, the sea ice of the Arctic Ocean and a few continental glaciers. Now, after 8000 years, these remnants are beginning to melt away. Alaska's spectacular Columbia Glacier has retreated twelve kilometres over the last twenty years; and in a few decades there will be no glaciers left in America's Glacier National Park. Glaciers such as these only contain enough water to alter the sea level by a matter of centimetres.

The Greenland ice cap, however, is a true remnant of the ice domes as big as whole continents, of the type mammoths would recognise. It contains enough water to raise sea levels globally by around seven metres. In the summer of 2002, it shrank by a record 1 million square kilometres—the largest decrease ever recorded. Two years later, in 2004, it was discovered that Greenland's glaciers were melting ten times faster than previously believed.

And the news keeps getting worse. In 2006 a report was published indicating that the Greenland glaciers are in fact melting twice as fast as we thought in 2004.

You might be surprised to learn that temperatures remain cold—indeed they are cooling—over the highest parts of both the Greenland and Antarctic ice domes. These are the only places on Earth where significant negative temperature trends are occurring. This is comforting, for a recent study has concluded that should the Greenland ice cap ever melt it would be impossible to regenerate it, even if our planet's atmospheric CO_2 was returned to pre-industrial levels.

The greatest extent of ice in the Northern Hemisphere is the sea ice covering the polar sea, and since 1979 its extent in summer has contracted by 20 per cent. And the remaining ice has greatly thinned. Measurements taken using submarines reveal that it is only 60 per cent as thick as it was four decades earlier. This huge melting, however, has no direct consequence for rising seas, any more than the melting ice cube in a drink raises the level of liquid in the glass.

This is because the Arctic ice cap is sea ice, nine-tenths of which is submerged, and when it melts it condenses into water in precisely the same proportion as it projected from the sea.

Only land ice, as it melts and runs into the sea, adds to sea levels.

Although the melting of sea ice has no direct effect, its indirect effects are important. At its current rate of decline, little if any of the Arctic ice cap will be left in summer by 2030, and this will significantly change the Earth's albedo.

Remember, one-third of the Sun's rays falling on Earth are reflected back to space. Ice, particularly at the Poles, is responsible for a lot of that albedo, for it reflects back into space up to 90 per cent of the sunlight hitting it.

Water, in contrast, is a poor reflector. When the Sun is overhead, water reflects a mere 5 to 10 per cent of light back to space. As you may have noticed while watching a sunset by the sea, however, the amount of light reflected off water increases as the Sun approaches the horizon. Replacing Arctic ice with a dark ocean will result in a lot more of the

Sun's rays being absorbed at Earth's surface and re-radiated as heat. This will create local warming which, in a classic example of a positive feedback loop, will hasten the melting of the remaining continental ice.

Over the past 150 years the oceans have risen by 10 to 20 centimetres, which amounts to 1.5 millimetres per year—around a hundredth as fast as your hair grows. Over the last decade of the twentieth century, the rate of sea level rise doubled to around three millimetres per year.

Scientists are concerned at the momentum of the rise, for the sea is the greatest single force on our planet. When movements within it reach a certain pace, all the effort of all the people on Earth can do nothing to slow it. The oceans, of course, are enormous when compared with the atmosphere, having 500 times the mass, and they are very dense.

So when we think of the atmosphere changing the oceans, we must imagine something like a VW Beetle pushing a tank down a slope. It takes effort to get the monster moving, but when it does begin to shift there's not much the Beetle can do to alter the tank's trajectory.

When our planet is heating it takes the surface layers of the oceans about three decades to absorb heat from the atmosphere, and a thousand years or more for this heat to reach the ocean depths. From the perspective of global warming, the oceans are still living in the 1970s.

Despite this, temperatures are rising at the surface of the oceans, and there is also a sharp rise in temperature at depth. There is nothing we can do to prevent this slow transfer of heat from air to sea, which is very bad news.

When most of us think of a rising sea, we imagine melting glaciers and ice caps pouring into the oceans. But oceans have another way to rise too. Over the past century much of the sea level rise has come from an expansion of the oceans, because warm water occupies more space than cold.

This thermal expansion of the oceans is expected to raise sea levels by 0.5 to 2 metres over the next 500 years.

Antarctica provides the most alarming news of melting ice. In February 2002 the Larsen B ice-shelf broke up over a matter of weeks. At 3250 square kilometres it was the size of Luxembourg. Scientists knew that the Antarctic Peninsula was warming more rapidly than almost anywhere else on Earth, but the speed and suddenness of Larsen B's collapse shocked many.

Why did Larsen B break up? Summer melting at both the top and the bottom of the ice sheet, brought about by warming of both the atmosphere and the ocean, had thinned it and fractured it with crevasses. But melting of the ice from below was the most important factor. While the Weddell Sea's deep waters, which flow past the ice, were still cold enough to kill a person in minutes, they had warmed by 0.32°C since 1972. This change was enough to initiate the melting.

Scientists are convinced that sometime this century the rest of the Larsen ice-shelf will break up, but by then our attention will be gripped by the fate of far greater ice-masses. The first is likely to be the Amundsen ice plain, an extensive area of sea ice off the coast of West Antarctica. NASA researchers have discovered that large sections of the ice plain are now so thin that they might float free of their 'anchors' on the ocean bed and collapse like Larsen B. The fatal moment for the Amundsen could be as soon as 2009.

By 2002 the glaciers feeding into the Amundsen had increased their rate of discharge to around 250 cubic kilometres of ice per year—enough to raise sea levels globally by 0.25 of a millimetre per annum. There is enough ice in the glaciers feeding into the Amundsen Sea to raise global sea levels by 1.3 metres.

The West Antarctic ice sheet is also tenuously anchored to the bottom of a shallow sea. It is one of the world's largest surviving expanses of sea ice. If it ever does detach from the sea floor, it would add 16 to 50 centimetres of sea level rise by 2100. Even worse, the glaciers feeding into it would accelerate, adding much more to sea levels. In all, the 3.8 million cubic kilometres of sea and glacial ice contained and held back by the West Antarctic ice sheet comprise enough water to raise global sea levels by six to seven metres.

There is one bright spot in all of this. The increased rainfall occurring at the Poles is expected to bring more snow to the high Antarctic ice cap, which may compensate for some of the ice being lost at the continent's margins, though how much and for how long is unknown.

Climate scientists are now debating whether humans have already tripped the switch that will create an ice-free Earth. If so, we have already committed our planet and ourselves to a rise in the level of the sea of around sixty-seven metres.

The bulk of the sea level rises will occur after 2050, but scientists are increasingly worried about large rises in the near future. In 2001 most were talking of a rise of a few tens of centimetres this century. By 2004 respectable scientists were predicting a rise of three to six metres over a century or two, while by 2006 Dr James Hansen, one of the most eminent climate scientists in the USA, was saying that a rise of 25 metres might occur within a few centuries.

The melting Poles might open a Northwest Passage to cargo vessels, but will there be any functioning ports to receive them?

Of all the free services that a stable climate offers us, none is taken for granted as often as a stable sea level. Just think about the city you live in, or any seaside town. Can you imagine how much effort and resources would be soaked up trying to protect property if the sea starts to rise rapidly? There would be no time or money left over for other pressing issues. If we do not act quickly to stabilise our climate, you may live long enough to see villages, suburbs and whole cities swallowed by the sea.

3

THE SCIENCE OF PREDICTION

SIXTEEN

MODEL WORLDS

The basic tool used in climate change prediction is a computer model of Earth's surface and the processes at work there. Scientists then vary the inputs, allowing them to see, for example, how our climate might respond to a doubling of CO_2 in the atmosphere, or how the ozone hole affects climate.

Today there are around ten different global computer models that simulate the way the atmosphere behaves now, and project how it will behave in the future. The most sophisticated of them are in England, California and Germany.

The Hadley Centre for Climate Prediction and Research in England has the appearance of the modern cathedral of climate change research. The new building, completed in late 2003, soars overhead, an elegant amalgam of glass and steel designed to minimise energy use and its impact on the environment. Here more than 120 researchers strive to reduce the uncertainty of projections by producing ever more sophisticated models that mimic the real world.

If our planet was a uniform black sphere, the Hadley people would have an easy task, for doubling the CO_2 in the

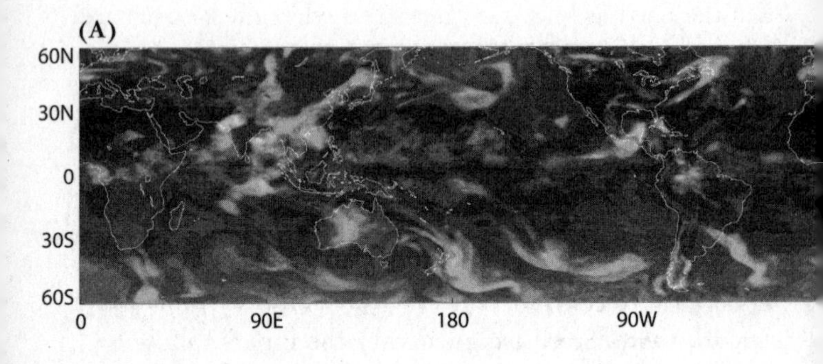

The weather for 1 July 1998. *(A)* is the Hadley Centre's computer simulation of world weather for the day; *(B)* is the actual weather as observed by satellite. The white arrows indicate cloud area that the computer failed to simulate, but otherwise the two images are very similar.

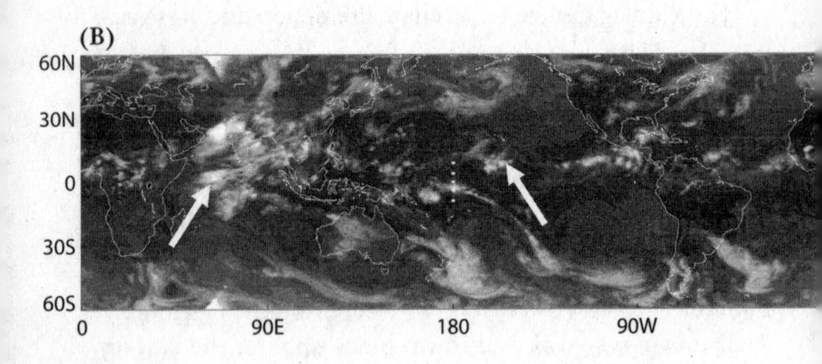

atmosphere would then raise its surface temperature by 1°C. But the Earth is blue, red, green and white, and its surface is bumpy. It is the white parts—much of which is cloud—that give the researchers headaches.

Clouds cloud the issue, so to speak, because no one has yet developed a theory of cloud formation and dissipation. Clouds can both trap heat and reflect sunlight back into space. This means they can, according to circumstances, either heat or cool.

So how good is the Hadley Centre's cloudy, computerised crystal ball at predicting Earth's future? There are four major tests that any global circulation model must pass before its predictions can be deemed credible.

- Is its physical basis consistent with the laws of physics—the conservation of mass, heat, moisture and so on?
- Can it accurately simulate the present climate?
- Can it simulate the day-to-day evolution of the weather systems that make up our climate?
- And finally, can the model simulate what is known of past climates?

The Hadley Centre's models pass all of these tests with a reasonable degree of accuracy, yet new discoveries in the real world are constantly forcing changes on these and other models.

For instance, we recently learned how human-induced climate change is altering sea level pressure. This is the first clear evidence of greenhouse gases directly affecting a meteorological factor other than temperature. Because the computer models hadn't taken account of this, we were underestimating the impact of climate change on storms in the North Atlantic.

Between the 1940s and 1970s, despite increasing greenhouse gas levels in the atmosphere, Earth's average surface temperature declined. Furthermore, early computer models predicted that, with the amount of CO_2 released into the atmosphere over the century, Earth should have been warming twice as much as it actually did.

Sceptics jumped on these inconsistencies to discredit both the models and the idea that CO_2 and other greenhouse gases caused temperatures to rise. Both discrepancies, it turned out, resulted from a previously overlooked factor—the very powerful influence on climate of minuscule particles that drift in the atmosphere.

Known as aerosols, they can be anything from dust ejected by volcanoes to the cocktail of deadly particles originating from the smokestacks of coal-fired power stations. Desert landscapes like those in the Sahel produce them in large quantities, and diesel engines, tyre rubber and fires are all important sources. Early computer models did not include aerosols in their calculations, in part because no one fully appreciated the extent to which human activities were increasing their number.

We now know that between one quarter and one half of all the aerosols in our atmosphere today are put there by human activity.

Aerosols can be very damaging to human health. They were the cause of significant mortality in seventeenth-century London when people were burning lots of coal. Even today aerosols generated by burning coal kill around 60,000 people annually in the US. Part of the reason is that coal acts like a sponge, soaking up mercury, uranium and other harmful minerals which are released when it is burned.

The state of South Australia is home to the world's largest uranium mine, yet its largest single point source of radiation is not the mine but a coal-fired power plant at Port Augusta. People worry about the radiation released from nuclear tests, but in the course of a year a single coal-fired power plant in Australia's Hunter Valley (and there are several such plants in the region) can release as much radiation into the atmosphere as the entire underground French nuclear testing program in the Pacific did. It's no real surprise that lung cancers commonly result: in the Hunter Valley lung cancer rates are a third higher than in nearby Sydney, despite the air pollution evident in the metropolis.

As a child I remember seeing No Spitting signs on the railway tunnel walls of my home city of Melbourne, and hearing stories of spittoons being used in my grandfather's day. When I travelled to China as an adult, and saw the inhabitants of grossly polluted cities such as Hefei hacking up foul congestion from their lungs, I realised that my forebears did not necessarily have worse hygiene habits than we do. They simply battled with a filthy atmosphere created by burning coal.

Scientists now think that the temperature decline of the 1940s to 1970s was caused by aerosols. One of them was sulphur dioxide which is released when low quality coal is burned. By the 1960s lakes and forests at high latitude in the Northern Hemisphere were dying. The trees were losing their needles, while the lakes were becoming crystal clear and emptied of life.

The cause was acid rain from the sulphur dioxide emissions of coal-burning power stations. Laws were passed to enforce the use of 'scrubbers' on coal-burning power plants in the industrialised world. These have been used since the 1970s and have dramatically reduced sulphur dioxide emissions.

There was, however, an unintended consequence. Aerosols of sulphate are most effective at reflecting sunlight back into space, and thus act powerfully to cool the planet. Because most aerosols last just a few weeks in the atmosphere (with sulphur dioxide degrading at the rate of 1–2 per cent per hour at normal humidity), the effect of installing scrubbers was immediate.

As the air cleared, global temperatures rose, driven by CO_2 released from those very same power stations. The experience was the perfect example of how everything on our planet is connected with everything else.

The 1991 eruption of Mt Pinatubo in the Philippines provided an exceptional test of the new computer models' capacity to predict the influence of aerosols. It ejected 20 million tonnes of sulphur dioxide into the atmosphere, and a group led by NASA scientist James Hansen forecast that the result would be around 0.3°C of global cooling—and this figure is *exactly* what was seen in the real world.

Among the most important and best supported of these models' predictions are that the Poles will warm more rapidly than the rest of the Earth; temperatures over the land will rise more rapidly than the global average; there will be more rain; and extreme weather events will increase in both frequency and intensity.

Changes will also be evident in the rhythms of the day, and nights will be warmer relative to days, for night is when Earth loses heat through the atmosphere to space. There will also be a trend towards the development of semi-permanent El Niño-like conditions.

We must now turn to the key uncertainty that remains in all models: will a doubling of CO_2, from pre-industrial levels of 280 to 560 parts per million, lead to a 2°C or 5°C increase in warming? After almost thirty years of hard work and profound technological advances we are still not sure about the answer to this question.

But many people would argue that we already know enough: even 2°C of warming would be catastrophic for large segments of humanity.

The largest study of climate change so far undertaken was published in 2005 by a team led from Oxford University. It was conducted by using the downtime on more than 90,000 personal computers, and it focused on the temperature implications of doubling CO_2 in the atmosphere. The average result was that this would lead to 3.4°C of warming. But there was an astonishingly wide range of possibilities—from between 1.9 and 11.2°C of warming, the higher end of which had not been predicted earlier.

As I read these results, I thought about an anomaly that had long niggled at me. At the end of the last ice age CO_2 levels increased by 100 parts per million, and Earth's average surface temperature rose by 5°C. Yet in most computer analyses, an increase in CO_2 almost three times as large is predicted to result in a temperature rise of only 3°C.

Of course, Milankovich cycles and extensive ice caps played a role, but scientists working on aerosols now think that they might have part of the answer. Direct measurement of the strength of sunlight at ground level, and worldwide records of evaporation rates (which are influenced primarily by sunlight) indicate that the amount of sunlight reaching the Earth's surface has declined significantly (up to 22 per cent in some areas) over the last three decades. The aerosols are blocking the sunlight.

This phenomenon is called global dimming, and there are two ways that it operates: aerosols such as soot increase the reflectivity of clouds, and the contrails of vapour left by jet aircraft create a persistent cloud cover. Soot particles change the reflective properties of clouds by encouraging many tiny water droplets rather than fewer, larger ones. These tiny water droplets allow clouds to reflect far more sunlight back into space than do larger drops.

The story with contrails is different. In 2001, in the three days following September 11 when terrorists destroyed the World Trade Center in New York, the entire US jet fleet was grounded. Over this time climatologists noted an unprecedented increase in daytime temperatures relative to night-time temperatures. This resulted from the additional sunlight reaching the ground in the absence of contrails.

If 100 parts per million of CO_2 really can raise surface temperature by 5°C, and if aerosols and contrails have counterbalanced this so that we have experienced only 0.63°C of warming, then their influence on climate must be enormously powerful. It is as if two great forces—both unleashed from the world's smokestacks—are tugging the climate in opposite directions, only CO_2 is slightly more powerful.

This leaves us with a grave problem, for particle pollution lasts only days or weeks, while CO_2 lasts a century.

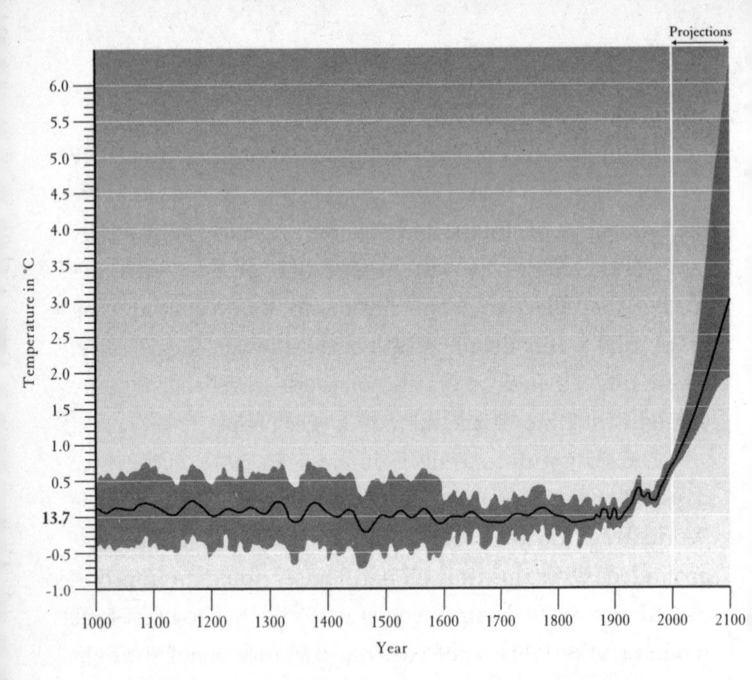

This graph, known as the 'hockey stick', shows trends in the average surface temperature of Earth from AD 1000 to 2100. Prior to 1900 this was 13.7°C. The grey area conveys uncertainty, which is reduced around 1850 when the thermometer grid was established. The projections on the right give a range of probable temperature increases to 2100.

If our understanding of global dimming is correct, then we only have one option. We must start to learn how to extract CO_2 from the atmosphere, and at the moment we don't know how to do that at all effectively.

One day we may be able to create artificial photosynthesis which would capture carbon out of the air but this belongs to a future we can at present only imagine.

One of the most fundamental human responses to any change is to ask what caused it. But Earth's climate system is so riddled with positive feedback loops that our usual concepts of cause and effect no longer hold. Consider that famous example from chaos theory, of the flutter of a butterfly's wing in the Amazon causing a cyclone in the Caribbean. To say simply that one thing has caused something else is an unhelpful way of thinking. Instead what we have are seemingly insignificant initial occurrences—such as an increase of atmospheric CO_2—that lead to runaway change.

A number of climate groups have produced computer-based projections for various regions of Earth and for time scales as short as a few decades. Here are three examples.

The Hadley Centre made predictions for the climate of the United Kingdom from the 2050s to the 2080s. They discovered that by 2050 human influences on the climate will have surpassed all natural influences.

They predict that snow cover will decrease by up to 80 per cent near the British coast, and up to 60 per cent in the Scottish highlands. Winter rainfall is predicted to increase by up to 35 per cent, with more intense rainfall events, while summer rainfall will decrease, and one summer out of three will be 'very dry'. An event akin to the severe summer of 1995 (which had seventeen days over 25°C and four days over 30°C) may recur twice per decade, while the great majority of years will be warmer than the record-breaking 1999. In 2006 the southeast of England had lapsed back into drought.

The changes felt over Europe will be more extreme than the increase in the global average. Indeed, a global rise in surface temperature of just 2°C would bring a temperature rise to all of Europe, Asia and the Americas of 4.5 degrees. For Britain, this means a more Mediterranean-like climate and, as some newspapers put it, 'the end of the English garden'. More important are the challenges it throws up for water security, flood planning and human health.

In 2003 and 2004 two further regional studies focused on climate impacts for California. They argued that global warming would bring much hotter summers to the state, as well as a depleted snowpack, threatening both water supplies and health. By the end of the century, heatwaves in Los Angeles would be two to seven times as deadly as today, and almost all of California's alpine forests would be lost. Already pikas (alpine relatives of the rabbit) are becoming extinct on isolated mountains throughout the west. Seven populations of around fifty have vanished in recent decades.

The third example focuses on the state of New South Wales, with predictions made by Australia's leading science research body, the CSIRO. It forecasts, in the decades to come, temperature increases across the state of between 0.2 and 2.1°C, while the number of cold spells and frosts will decrease. The number of days above 40°C will increase, as may winter and spring droughts, extreme rainfall events and wind speeds.

The gas is already in the air and right now we have no way of getting it out. Whatever the accuracy of these reports, one thing is certain: the course of climate change is set for at least the next few decades.

SEVENTEEN

DANGER AHEAD

The full impact of the greenhouse gases already in the atmosphere will not be felt until around 2050. If greenhouse gas emissions stopped immediately, that date is when Earth would reach a new stable state, with a new climate. That's because of the long life of CO_2 in the atmosphere. Researchers call this 'the commitment': change we are yet to feel but can't prevent.

Much of the CO_2 released when people stoked their coal-fuelled stoves in the aftermath of World War I is still warming our planet today. Most of the damage was done starting from the 1950s, when people drove about in their fin-tailed Chevrolets and powered their labour-saving household appliances from inefficient coal-burning power stations.

It is the baby-boomer generation born in the years following World War II that is most to blame: half of the energy generated since the Industrial Revolution has been consumed in just the last twenty years.

It's easy to condemn this extravagance, but we must remember that until recently nobody had the slightest idea that their auto emissions or electric vacuum cleaner would have an impact on their children and grandchildren.

But now we know. The true cost of your family's four-wheel-drives, air conditioners, electric hot water services, clothes dryers and refrigerators is increasingly evident to all. In many developed nations we are three times as affluent on average as people were a few decades ago, and therefore we are able to bear the cost of changing our ways.

Our commitment—the climate change we can no longer prevent—is influenced by a number of factors:

- the CO_2 we have already released;
- the positive feedback loops that amplify climate change;
- global dimming;
- the speed at which human economies can decarbonise themselves.

The first—existing greenhouse gas volumes—is known and cannot be changed.

The second and third—positive feedback loops and global dimming—are still being explored by scientists.

And the fourth—the rate at which we humans can reduce our emissions—is being argued over right now in parliaments and boardrooms around the world. It and global dimming are the only impacts over which we have control.

Scientists say that a 70 per cent reduction in CO_2 emissions from 1990 levels by the middle of the twenty-first century is required to stabilise Earth's climate. This would result in an atmosphere with 450 parts per million CO_2—

remember it's now at 380 parts per million. Our global climate would stabilise by around 2100 at a temperature at least 1.1°C higher than the present, with some regions warming by as much as 5°C.

The European nations are talking of emissions cuts on this scale but, given the resistance of the coal industry and the policies of the Bush and Howard administrations in the US and Australia, this may not be an achievable target. A more realistic scenario may be 550 parts per million of atmospheric CO_2—double the pre-industrial level. This would result in climatic stabilisation centuries from now, and an increase in global temperature of around 3°C this century.

But remember that even this depends on good luck. The level of greenhouse gases already in the atmosphere may trigger positive feedback loops with the potential to cause change we can't control.

It's too late to avoid altering our world, but we still have time to avoid disaster and to reduce the probability of dangerous climate change.

Perhaps a more useful way of looking at the problem is to work out *rates* of change that are dangerous. After all, life is flexible, and if given sufficient time it can adapt to the most extreme conditions. But if change occurs too quickly, plants and animals won't have time to adapt. If we think about it this way, warming rates above 0.1°C per decade are likely to cause grave damage to our ecosystems. Similarly, rates of sea level rise above two centimetres per decade would be dangerous, as would a rise of five centimetres overall.

But the question of what constitutes dangerous climate change raises another question—dangerous to whom? For the Inuit in the Arctic a damaging threshold has already been crossed. Their primary food sources of caribou and seal are now difficult to find as a result of climate change and their villages are under threat.

When we consider the fate of the entire planet, there can be no illusions about what is at stake. Earth's average temperature is around 15°C, and whether we allow it to rise by a single degree, or 3°C, will decide the fate of hundreds of thousands of species, and billions of people.

EIGHTEEN

RETREATING UP THE MOUNTAINS

Neither the snows of Mount Kilimanjaro in Africa nor the glaciers of New Guinea can survive current levels of CO_2 for more than a couple of decades. And below those icy realms, every habitat, each of which has its own unique species, is climbing upwards.

Nothing in climate prediction is more certain than the extinction of many of the world's mountain-dwelling species.

We know that our planet must heat by 1.1 degrees this century come what may. Continuing the way we are will commit us to a 3°C increase in temperature. The very highest peak in New Guinea—Puncak Jaya—is just under 5000 metres. A rise of 3°C would push the last of New Guinea's alpine habitat off its summit. Indeed, given such extreme changes, there are few mountains anywhere on Earth high enough to provide an alpine refuge.

Waking in the crisp air on a New Guinea mountaintop and seeing delicate spiderwebs strung between tree-ferns, glittering with dew, is an experience to cherish. In the slanting light the dominant colours of these open, equatorial meadows are bronze and brilliant green, interspersed with the flamboyant red, orange and white flowers of tussock-like rhododendrons and orchids. Around your feet in the mossy soil are the scratchings of the long-beaked echidna—at a metre in length it's the largest egg-laying mammal on Earth. And if you look you'll see the burrows of the alpine woolly rat. It's also a giant, measuring almost a metre from nose to tail-tip.

Dingiso. The amazing black and white tree-kangaroo lives in alpine environments in West Papua. In 1994, my colleagues and I 'discovered' it. Tragically, the labrador-sized marsupial may become a victim of climate change.

At dawn the air is full of birdsong, for these mountains are the retreat of birds of paradise, parrots, and hordes of honeyeaters that flock to the flower-filled bushes. Towards mid-morning, from the scattered marshes you'll hear *Oooh, ooh*, which you might think (as I did) sounds like your favourite aunt, tipsy after Christmas dinner. But here is a tiny rose-pink frog—no bigger than a child's thumbnail—and so new to science that it hasn't yet been named.

Every high tropical mountain on Earth has an equivalent alpine habitat, and below them are mountain forests that are even richer in life. The world's mountain ranges nurture a staggering variety of life—from iconic species such as pandas and mountain gorillas to humble lichens and insects. Although alpine habitats make up a mere 3 per cent of the surface of the Earth, they are home to over 10,000 plant species, along with countless insects and larger animals.

Over the course of the twentieth century, mountain-dwelling species have each decade withdrawn on average 6.1 metres up the slopes of their homes. They did this because conditions lower down became too hot or dry, or because new species arrived with which they could not compete.

The rainforest-clad mountains of northeast Queensland are centred on the Atherton Tablelands west of Cairns, and cover 10,000 square kilometres. Despite their small size they are arguably the most important habitat in all of Australia, because they are home to plants and animals that are survivors from the cooler, moister Australia of 20 million years ago.

In 1988 the rainforests were listed as Australia's first World Heritage Area. Tourists now flock there, and one of the most popular activities is a night walk, when an abundance of marsupials can be seen at close range by spotlight. In some places the forest is alive with grunts, squeals and rustlings.

High up in the tallest trees you'll hear lemuroid ringtail possums leaping from branch to branch. They are living fossils, related to the majestic metre-long greater glider of the eucalypt forests. Lemuroids lack a gliding membrane, but are extraordinary leapers whose noisy crashing through the canopy is one of the most constant noises at night.

Lower down in the trees you might see the green ringtail possum with her large young. They are fussy eaters and the young stays with its mother until it's almost adult-sized so it can learn which leaves are best. The possums haunt the mountain summits because if they spend even four or five hours in temperatures of 30°C or more they will die. These temperatures are an almost daily event in the surrounding lowlands.

Sixty-five species of birds, mammals, frogs and reptiles are unique to the region and none can tolerate warmer conditions. They include the golden bowerbird, the Bloomfield nursery frog and Lumholtz's tree-kangaroo.

Higher CO_2 levels affect plant growth. Plants grown

experimentally in CO_2-enriched environments tend to have reduced nutritional value and tougher leaves. This change alone is predicted to reduce possum numbers. As species retreat upwards the poor soils that dominate on the summits will further reduce the nutritional value of their food. If this were not enough, rainfall variability is likely to increase, with droughts becoming more severe. The cloud layer, which now provides 40 per cent of the water that nourishes the mountain forests, will rise, exposing the forests to more sunlight and thus evaporation. All of this adds up to a catastrophic impact.

With the inevitable rise in temperature of just 1°C, at least one unique wet tropics species—the Thornton Peak nursery frog—will become extinct. With a 2°C increase the wet tropics ecosystems will begin to unravel. At a 3.5°C increase, around half of the sixty-five species of animals unique to the wet tropics will have vanished, while the rest will become restricted to tenuous habitats of less than 10 per cent of their original size. In effect their populations will be non-viable, their extinction only a matter of time.

The implications for the future of Australia's biodiversity are enormous. For example, the bunya pine—a relative of the monkey puzzle tree and the oldest species in an ancient family—is restricted to two mountain ranges. This species, or something like it, has been around since the Jurassic Era some 230 million years ago. Its loss would be calamitous. As would the loss of orchids, ferns and lichens

Lumholtz's tree-kangaroo is one of about 10,000 species of plants and animals that are unique to the mountain forests of northeastern Queensland. With a 3.5°C increase in temperature, their habitat will cease to exist.

or the invertebrates—those legions of worms, beetles and other flying and crawling things found in their tens of thousands. Can you imagine Australia without the Atherton rainforests and the Great Barrier Reef too?

The impending destruction of Australia's wet tropics rainforests is a biological disaster on the horizon, and the generations responsible will be cursed by those who come after.

What will they tell their children if their air conditioners and four-wheel-drives cost them the nation's natural jewels?

Throughout the world, every continent and many islands have mountain ranges that are the last refuge of species of remarkable beauty and diversity. And we stand to lose it all, from gorillas to pandas to the wonderfully named vegetable sheep, tussocky plants which grow only in the alpine areas of New Zealand. There is only one way to save them. We must act to stop the problem at its cause—the emission of CO_2 and other greenhouse gases.

There is, surprisingly, one group of species that will benefit enormously from this aspect of climate change. These are the parasites that cause the four strains of malaria. As rainfall increases the mosquitoes that carry the parasite will spread, the malarial season will lengthen and the disease will proliferate. From Mexico City to Papua New Guinea's Mt Hagen, the mountain valleys of the world support human populations in high densities. And, where population density is not too great, they are glorious places in which disease is rare.

Just below these communities—in the case of New Guinea at around 1400 metres—are great forests where no one lives because malaria thrives there. In the near future, global warming will allow the malarial parasite and its vector the *Anopheles* mosquito to enter those high mountain valleys. There they will find tens of thousands of people without any resistance to the disease.

NINETEEN

HOW CAN THEY KEEP ON MOVING?

Species have survived climate change in the past because mountains have been tall enough, continents big enough and the change slow enough for them to find new habitats which suit them. The key to survival in the twenty-first century will be for animals and plants to keep on moving. But how will they manage this?

For instance, should Australia's temperature rise by only 3°C over this century, half of Australia's *Eucalyptus* species would grow outside their current temperature zone. If they are to survive they must migrate, yet numerous barriers, including the Southern Ocean and human-modified landscapes, stand in the way.

South Africa's succulent Karoo flora comprises some 2500 species of plants found nowhere else. It is the richest arid-zone flora on Earth, and is renowned for the beauty of its spring flowers, which depend on marginal winter rainfall. As the climate changes, where can this vegetation go? To the south and east—the direction which climate change will drive it—lie the Cape Fold Mountains, whose topography and soils are unsuitable for Karoo plants. Computer models indicate that 99 per cent of the Karoo will have vanished by 2050.

To the south of the Cape Fold Mountains is the scrubland of the fabulous fynbos, one of six floral kingdoms on Earth, and the most diverse plant community to be found outside the rainforests. The plants are little more than knee-high, but their form is extraordinary. Rushes bear brilliant bell-shaped flowers, whose nectar is sipped by brightly coloured 'humming flies' with two-centimetre-long siphons which reach deep into the bells. Rocky slopes are adorned with bushy king proteas studded with saucer-sized, pink star-flowers. The profusion of pea flowers, daisy-like forms and iris flora seems endless.

Hemmed in by ocean at the southern tip of the continent, the fynbos is a natural paradise, but it is cornered. As the Earth warms it will lose over half of its extent by 2050.

The diverse heathlands of Australia's southwest comprise over 4000 species of flowering plants. With just half a degree of additional warming, the fifteen species of mammals and frogs unique to the region would be restricted to tiny habitats, or would become extinct. Yet we already know that half a degree of warming is inevitable.

Global warming could not have come at a worse time for biodiversity. In the past, when abrupt shifts of climate occurred, trees, birds, insects would migrate the length of continents. In the modern world, with its 6.5 billion humans, such movements are not possible. Today, most biodiversity is restricted to national parks and forests.

The wintering habitat for migratory shore birds in North America will be reduced because of drying trends, rising seas and increased storm surge. Warming streams mean that salmon will diminish in number, while in the North Atlantic commercially valuable fish are already following the cold water downwards and northwards.

By 2005–6 many of these changes were already evident. The Frazer River in British Columbia, which is one of the world's most important salmon spawning streams, had been fatally hot for salmon during six of the prior fifteen years. Warming sea water off the coast of British Columbia had also led to a collapse in population of the small crustaceans that form the basis of the food chain. This in turn led to a lack of fish and other marine life, which had severe impacts on the larger creatures.

Cassulet's auklet is a small seabird whose major breeding colony is on Triangle Island, British Columbia. In 2005 one million birds gathered to breed there, but such was the shortage of food that not a single chick was known to have survived. Failure of breeding on this scale has not been recorded in the many decades that the birds have been monitored.

Whale watchers noticed that the humpback whales migrating down the coast of British Columbia towards Hawaii were 'misshapen' through starvation. Shortly after, divers swimming with the whales in their wintering grounds off Hawaii, watched in puzzlement as the whales tried to feed in the virtually nutrient-free waters. For a biologist seeing all of this, the situation seems uncannily reminiscent of the situation the golden toad found itself in by 1987.

Mexico's fauna will be squeezed by heat and extreme weather events, resulting in many extinctions. These factors have also led botanists to declare that one third of Europe's plant species face severe risks.

On smaller landmasses the situation is even bleaker. Many Pacific island birds will be pushed beyond their limits, and there will be extinctions in all forms of life, from trees to insects. Kruger National Park in South Africa is nearly the size of Israel, yet it still stands to lose two-thirds of its species.

Imagine what would happen if Washington's climate were more like Miami, or Sydney's more like Cairns. Try to think what this change would mean for the forests, birds and animals of the region where *you* live. You'll begin to see the bigger picture.

The bigger picture includes the deepest parts of our oceans. Even the fish that live there can teach us about climate change. When marine biologists haul up bizarre creatures from the depths, the animals are already dying. Black, toothy bodies of deep-sea anglers lie inert, their luminescence slowing to a flicker. Predators such as the stoplight loosejaw grow pale and vomit up their last meal, often a fish larger than themselves. Within minutes they stop moving and their eyes glaze over.

It's the change in pressure that killed them, scientists used to say. At the depths where these creatures live, the force of the kilometres-high column of water overhead is so intense that a submarine would buckle in an instant. As proof of this idea experts pointed to those few deep-sea fishes that have swim-bladders. They reach the surface grossly distorted, their air-sacs swollen with expanding gas and their bodies stretched to bursting. Despite such gruesome evidence, we now know differently.

In your imagination, grit your teeth and pick up a hairy seadevil that has just emerged from a depth of three kilometres. Trust me, it is the ugliest of all fishes. Then toss its black, sack-like and filament-covered body into a bucket of icy sea water. Now step back.

Within minutes vitality will return to its frame, its great fang-studded jaws will snap, and the filament-clotted 'fishing rod' that protrudes from between its eyes will flicker. Its life, you see, is not threatened by pressure but by warmth. In the deep ocean water where it lives temperatures hover near zero. Waters that would chill us to death in minutes are fatally warm to these fish.

The structure of the world's oceans is critical to our climate. There are three layers, separated by their temperature. The top 100 metres or so vary enormously. Near the

Poles it can be below zero, while at the equator it can exceed 30°C.

As you descend below this familiar, light-filled world, so does the mercury in the thermometer. At around a kilometre down we have reached the world's deep ocean water, and from bottom to top it's remarkably stable in temperature. It varies between –0.5°C (it can be below freezing and not turn to ice because of the salt) and 4°C. Most of the water in this lightless realm is exported from Antarctica, where it has been chilled to near freezing point by submarine currents.

If we continue to warm our planet, the astonishing residents of the deep will eventually be heated to death. But they also face another danger, which will first manifest itself where the icy water of the deep ocean comes to the surface near the Poles. As the oceans absorb more CO_2 they become acid. The CO_2 reacts with the carbonate present in the oceans, and the carbonate can drop below the level at which it can be used by shell-forming animals, like oysters, crabs and prawns. As it becomes impossible for them to maintain their protective covers they die.

Scientists used to think that increasing acidity would not become a big problem for centuries. Then in 2005 a new study indicated that the situation had got much worse. Dangerously acidic waters may develop in the next few decades in vulnerable regions such as the northern Pacific Ocean. This is a truly frightening possibility, for this acidity would severely damage the ocean ecosystem and its ability to produce food for us.

If you would like future generations to know the taste of prawns and oysters, we all need to limit CO_2 emissions now.

The problem of acidity, incidentally, has nothing to do with global warming. So it should be deeply concerning even for those who deny the reality of climate change.

If we do act now, we can save many species whether they live in the oceans or on land. Some scientists believe that, at the lowest amount of inevitable global warming—between 0.8 and 1.7°C—around 18 per cent of all the species alive today are doomed. That's one in five of all species.

At the mid-range projections—1.8 and 2.0°C—around a quarter of all species will vanish, while at the high range of predicted temperature rises (over 2°C) over a third of species will become extinct.

Believe it or not, this is the good news; in these projections it is assumed that the species can migrate. But what chance does a protea have of making it across the populated coastal plain of South Africa's Cape Province, or a lion-tamarin monkey crossing the agricultural fields that have all but obliterated the Brazilian Atlantic rainforests?

For species that cannot migrate the likelihood of extinction is roughly doubled. This means that, at the high range of predicted temperatures, over half (58 per cent) of all species will be committed to extinction.

It appears that, without human assistance, at least one out of every five living things on this planet is doomed to vanish by the existing levels of greenhouse gases. If we don't make changes now, in all likelihood three out of every five species will become extinct at the dawn of the next century.

The World Wildlife Fund, the Sir Peter Scott Trust and the Nature Conservancy have all worked for decades to save, in real terms, relatively few species. Now thousands might be swept away unless greenhouse gas emissions are reduced.

We must remember this. If we act now we can save at least four out of every five species.

TWENTY

THE THREE TIPPING POINTS

Scientists are aware of three big tipping points for Earth's climate: a slowing or collapse of the Gulf Stream; the death of the Amazon rainforests; and the explosive release of methane from the sea floor.

All three occur in the virtual worlds of the computer models, and there is some geological evidence for all having happened at various times in Earth history. Given the current rate and direction of change, one, two or perhaps all three may take place this century. So what leads to these sudden shifts, what are the warning signs, and how might they affect us?

SCENARIO 1
Collapse of the Gulf Stream

The importance of the Gulf Stream to the Atlantic rim countries is enormous. In 2003 the Pentagon commissioned a report outlining the implications for US national security should the Gulf Stream collapse. The purpose of the report was, its authors said, 'to imagine the unthinkable'.

In their scenario the Gulf Stream slows as a result of fresh water from melting ice pouring into the North Atlantic. The planet continues to warm until 2010, but then a dramatic shift will occur—a 'magic gate' that will abruptly alter the world's climate.

The Pentagon 'weather report' for 2010 predicts persistent drought over critical agricultural regions, and a plunge in average temperatures of more than 3°C for Europe, just under 3°C for North America, and 2°C increases for Australia, South America and southern Africa.

The report predicts that nations won't co-operate with each other in the face of the disaster: mass starvation would be followed by mass emigration. Regions as diverse as Scandinavia, Bangladesh and the Caribbean would become incapable of supporting their populations. New political alliances would be forged in a scramble for resources. And war would be likely.

By 2010–20, with water supplies and energy reserves strained, Australia and the US would focus increasingly on border protection to keep out the migrating hordes from Asia and the Caribbean. The European Union, the report says, may go one of two ways—either it would focus on border protection (to keep out those homeless Scandinavians, among others), or be driven to collapse and chaos by internal squabbling.

In 2004 the Hollywood disaster movie *The Day after*

Tomorrow also imagined the consequences of the shutdown of the Gulf Stream. For dramatic effect the time-lines for the collapse are greatly compressed, and the changes are far grander even than those imagined in the Pentagon report.

Scientists meanwhile have been working at understanding the consequences for biodiversity if the Gulf Stream were to collapse. They *are* catastrophic. If the currents will no longer carry oxygen into deeper waters, biological productivity in the North Atlantic will fall by 50 per cent, and oceanic productivity worldwide will decrease by over 20 per cent.

So what are the chances of the Gulf Stream shutting down this century? What would be the warning signs?

The Gulf Stream is the fastest ocean current in the world, and it is complex, spreading out into a series of spirals and sub-currents as its waters move northward. The volume of water in its flow is simply stupendous. You will recall that ocean currents are measured in Sverdrups, and one Sverdrup equals a flow of 1 million cubic metres of water per second. Overall, the flow rate of the Gulf Stream is around 100 Sverdrups, which is 100 times as great as that of the Amazon.

In its northern section the Gulf Stream is far warmer than the waters that surround it. Between the Faeroe Islands in Denmark and Great Britain, for example, the Gulf Stream is a balmy 8°C, yet the surrounding waters are at zero. The source of the Gulf Stream's heat is the tropical sunlight falling on the mid-Atlantic, and the current is a highly efficient way to transport it, for one cubic metre of water can warm 3000 cubic metres of air.

In the North Atlantic, where the Gulf Stream releases its heat, it warms Europe's climate as much as if the continent's sunlight were increased by a third.

As the waters of the Gulf Stream yield their heat they sink, forming a great mid-oceanic waterfall. This waterfall is the powerhouse of the ocean currents of the entire planet, but history shows us that it has been interrupted many times.

Fresh water disrupts the Gulf Stream because it dilutes its saltiness, preventing it from sinking and thus disrupting the circulation of the oceans worldwide. Several Sverdrups or more of freshwater flow is needed. If the frozen north melted it could achieve that liquid potential, and to this we must add the increasing rainfall across the region.

The tropical Atlantic is becoming saltier at all depths, while the North and South polar Atlantic are becoming fresher. The change is due to increased evaporation near the equator and enhanced rainfall near the Poles. When similar changes were observed in other oceans, scientists realised that something—most probably climate change—had accelerated evaporation and rainfall rates over the oceans by 5 to 10 per cent.

This increasing tropical saltiness could lead to a temporary quickening of the Gulf Stream before its shutdown. Extra heat would be transferred to the Poles, which would melt more ice until enough fresh water could flow into the North Atlantic, collapsing the system altogether.

How fast might it happen? Ice-cores from Greenland indicate that, as the Gulf Stream slowed in the past, the island experienced a massive 10°C drop in temperature in as little as a decade. Presumably, similarly rapid changes were also felt over Europe, although no detailed record of climate has survived to tell of it.

It is possible, if the Gulf Stream were to slow, that extreme falls in temperature could be felt over Europe and North America within a couple of winters.

When is such an event likely to happen? Some climatologists think they are already seeing early signs of a shutdown. Not all agree—scientists at the Hadley Centre in England rate the chance of major disruption to the Gulf Stream this century at 5 per cent or less. Their main concern is an event in the Amazon that could be even more catastrophic.

SCENARIO 2

Collapse of the Amazon rainforests

One of the Hadley Centre's computer models is known as TRIFFID (Top-down Representation of Interactive Foliage and Flora Including Dynamics). It suggests that, as the concentration of CO_2 in the atmosphere increases, plants—particularly in the Amazon—start behaving in unusual ways.

The plants of the Amazon effectively create their own rainfall—the volume of water they transpire is so vast that it forms clouds whose moisture falls as rain, only to be transpired again and again.

But CO_2 does odd things to plant transpiration. Plants, of course, generally don't wish to lose their water vapour, as they have gone to some trouble to convey it from their roots to their leaves. Inevitably they do lose some whenever they open the breathing holes (stomata) in their leaves. They open their stomata to gain CO_2 from the atmosphere, and they will keep them open only as long as required.

Thus, as CO_2 levels increase, the plants of the Amazonian rainforest will keep their stomata open for briefer periods, and transpiration will be reduced. And with less transpiration there will be less rain.

TRIFFID indicates that, by around 2100, levels of CO_2 will have increased to the point that Amazonian rainfall will reduce dramatically, with 20 per cent of that decline attributable to closed stomata. The rest of the decline, the model predicts, will be due to a persistent drought that will develop as our globe warms.

The current average rainfall in the Amazon of 5 millimetres per day will decline to 2 millimetres per day by 2100, while in the northeast it will fall to almost zero. These conditions, combined with an average rise in temperature of 5.5°C, will make the collapse of the Amazonian rainforest inevitable. A small change in temperature is capable of turning soils from absorbers of CO_2 to large-scale producers. As the soil warms, decomposition accelerates and lots of CO_2 is released. This is a classic example of a positive feedback, where increasing temperature leads directly to a vast increase in CO_2 in the atmosphere, which further increases temperature. With the loss of the rainforest canopy, soils would heat and decompose more rapidly, which would lead to the release of yet more CO_2.

This means a massive disruption of the carbon cycle. The storage of carbon in living vegetation would fall by 35 gigatonnes, and in soil by 150 gigatonnes. These are huge figures—totalling around 8 per cent of all carbon stored in the world's vegetation and soils!

The outcome of this series of positive feedback loops is that by 2100 the Earth's atmosphere would have close to 1000 parts per million of CO_2. Remember our current level is 380 parts per million and we need to act now to stop it reaching 550 parts per million.

This modelling experiment predicts devastation in the Amazon Basin. Temperatures rise by 10°C. Most of the tree-cover is replaced by grasses, shrubs, or at best a savannah studded with the odd tree. Large areas become so hot and blighted that they cannot support even this reduced vegetation, and turn into barren desert.

When might all of this happen? If the model is correct we would start to see signs of Amazonian rainforest collapse around 2040.

By the end of the century the process would be complete. Half of the deforested region will turn to grass, the other half to desert.

What is so terrifying about this scenario is that climate change in the Amazon would itself hasten further runaway global climate change.

SCENARIO 3
Methane release from the sea floor

Clathrates is the Latin word for 'caged' and the name refers to the way ice crystals trap molecules of methane. Clathrates are also known as the 'ice that burns'. They contain lots of gas under high pressure, which is why pieces of the icy substance hiss, pop and, if lighted, burn when brought to the surface.

Massive volumes of clathrates lie buried in the seabed right round the world—perhaps twice as much in energy terms as all other fossil fuels combined. The clathrates in the seabed are kept solid only by the pressure of the cold overlying water. There are masses of clathrates in the Arctic Ocean, where temperatures are sufficiently low, even near the surface, to keep them stable.

It's illustrative of the endless ingenuity of life that some marine worms survive by feeding on the methane in clathrates. They live in burrows within the icy matrix, which they mine for their energy requirements. There are between 10,000 and 42,000 trillion cubic metres of the stuff scattered around the ocean floor, compared with the 368 trillion cubic metres of recoverable natural gas in the world. It's not surprising that both worms and the fossil fuel industry can see a future in this weird material.

If pressure on the clathrates were ever relieved, or the temperature of the deep or Arctic oceans were to increase, colossal amounts of methane could be released. Palaeontologists are now beginning to suspect that the unleashing of the clathrates may have been responsible for the biggest extinction event of all time 245 million years ago.

At that time around nine out of ten species living on Earth became extinct. Known as the Permo-Triassic extinction event, it destroyed early mammal-like creatures, thus opening the way for the dominance of the dinosaurs.

Many people think the cause of the extinction may have been a massive outpouring of lava, CO_2 and sulphur dioxide from the Siberian Trap volcanoes (the largest known flood basalt area). This would have led to an initial rise in global average temperature of about 6°C and widespread acid rain, which would have released yet more carbon. The increasing temperature then triggered the release of huge volumes of methane from the tundra and from clathrates on the sea floor. The explosive power to change climate would have been beyond imagination.

Two of these scenarios—the Amazonian die-back and the release of the clathrates—involve positive feedback loops, where changes build on each other to produce even greater changes. But there's one other positive feedback loop that's already occurring, and may be the trigger for further change.

Throughout our history we have engaged in a constant battle to maintain a comfortable body temperature, which has been very costly in terms of time and energy. Just think of the hundreds of slight shifts in body position we make every day and night, and the taking off and putting on of overcoats and hats. Purchasing a house, our greatest personal expense, is primarily about regulating our local climate. In the US, 55 per cent of the total domestic energy budget is devoted to home heating and air conditioning. Home heating alone costs Americans US$44 billion per year.

As our world becomes more uncomfortable because of climate change, the demand for air conditioning will increase. In fact, during heatwaves it could mean the difference between life and death. But, unless we change how we create electricity, that demand for air conditioning will be met by burning more fossil fuels, which is a powerful positive feedback loop.

As global warming speeds up we will huddle at home clutching the remote of our climate control system, releasing ever more greenhouse gases. There is already a huge demand for air conditioners in countries such as the US and Australia where, until recently, construction codes for houses have been appallingly lax in regard to energy use.

Will we, in order to cool our homes, end up cooking our planet? Will air conditioning be one of the causes of the collapse of the Amazon or the interruption of the Gulf Stream?

TWENTY-ONE

THE END OF CIVILISATION?

Our civilisation is built on two foundations: our ability to grow enough food to support a large number of people who are engaged in tasks other than growing food; and our ability to live in groups large enough to support great institutions such as our parliaments, our courts, and our schools and universities.

These clusters are of course cities, and it is from their inhabitants, the citizens, that the word civilisation is derived.

Today, very large cities lie at the heart of our global society. They might be enormous but they are fragile entities and need to be subsidised from outside to receive their basic needs—food, water and power.

Our cities are like rainforests in their complexity.

In city life, almost every job is specialised. Being a mere 'secretary' will no longer do—one must instead be a legal or medical secretary, or some such. And a doctor is best served by becoming a sports specialist, a proctologist, or an expert on aged care. This is the equivalent, in human terms, of

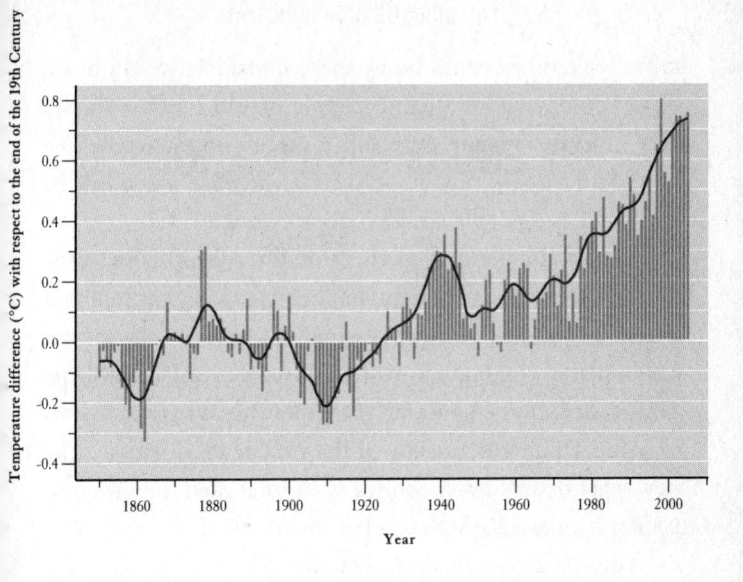

Average global temperatures from 1850 to 2005. Five of the six hottest years on record have occurred since 2000.

being a rainforest species like a *matanim* cuscus or a golden toad. Only in a rainforest is there a supply of energy and moisture that is big enough and regular enough to feed such specialised creatures.

As we have seen, if we cut off the water or sunlight to a rainforest for even a brief period it is liable to collapse and its specialist species become extinct. Now, let's perform a mental experiment. Think of a city you are familiar with and imagine what it would be like if its citizens woke one morning to discover that no fresh water came from their taps. No clothes could be washed, no toilets would flush, filth would accumulate and people would become thirsty very quickly. Imagine the result if petrol supplies came to a halt. Food could not be delivered, garbage wouldn't be removed, and people couldn't get to work.

Could climate change threaten the resources required by cities to survive? Physicist Stephen Hawking has said that within a thousand years increasing CO_2 will boil the surface of our planet and humans will need to seek refuge elsewhere. This is an extreme opinion. More mainstream are the views of Jared Diamond, author of the bestseller *Collapse*. The key reason why even complex, literate societies such as Central America's Maya failed, he found, is because they used up too many of their resources.

A rapid shift to another kind of climate could place a similar stress on our global society, for it would alter the location of sources of water and food, as well as their volume.

Climate change can intensify climate variability, making it difficult to predict how the weather will behave in the short and long terms. Australia provides a good example of the relationship between climate variability and human population size. It is unique among the larger nations in consisting of either very small settlements or large cities. There are almost no middle-sized towns of the kind that predominate elsewhere in the world. This is a consequence of the cycle of drought and flood that characterises its climate.

Small regional population centres in Australia have survived because they can endure drought. Large cities have also survived because they are integrated into the global economy. Middle-sized towns, however, are vulnerable. Typically, as a drought progresses, the farm machinery dealership and automotive dealership close down. Then the pharmacist, the bookseller and the banks leave. When the drought finally breaks and people have cash again, these businesses do not return. Instead people travel to larger centres to buy what they need, and in time end up moving there.

Australia is the most urbanised nation on Earth: a higher percentage of its population lives in cities than anywhere else in the world.

Australia shows that climate variability has in fact encouraged the formation of cities. But the only reason that Australia's cities are refuges from climate variability is that they draw their resources from an area broader than that affected by the continent's droughts and floods.

Climate change, however, is global: the entire Earth is affected by climate shifts and extreme weather events of increasing power, and Australian cities will not be exempt from this.

Water will be the first of the critical resources to be affected, for it is heavy, commands a low price and is unprofitable to transport long distances. Most cities source their water locally, in areas small enough for mild climate change to have an impact. Food such as grains, in contrast, is easily transported and is often sourced from afar, which means that only truly global disruptions would cause shortages in the world's cities.

For the past ten years, droughts and unusually hot summers have caused world grain yields to fall or stagnate, during which time the number of extra mouths humanity must feed has grown by around 800 million. So far, we have been able to cope with these relatively modest impacts from climate change.

When it comes to climate change, cities are more like plants than animals: they are immobile, and depend on intricate networks to supply the water, food and energy that they require. We should worry that whole forests are already dying as a result of climate change, for cities will also begin to die when their networks can no longer supply the essentials. The causes could be the repeated battering of extreme weather events, rising seas and storm surges, extreme cold or heat, water deprivation or flood, or even disease.

And we forget that our global civilisation is knit together by sea trade, and sea trade depends upon ports that could be rendered non-functional by a rising ocean.

Could the day ever come when the taps run dry, when power, food or fuel is no longer available in many of the world's cities?

If we were to experience an abrupt climate shift, it is possible that a near-eternal, dreary winter would descend on the cities of Europe and eastern North America, killing crops, and freezing ports, roads and human bodies. Or perhaps extreme heat, brought on by a vast exhalation of CO_2 or methane, will destroy the productivity of oceans and land alike.

Humanity, of course, would survive such a collapse, for people will persist in smaller, more robust communities such as villages and farms—situations that resemble temperate deciduous forests rather than rainforests. Towns have relatively few people just as temperate forests have relatively few species, and the inhabitants of both are hardy and multi-skilled. Just think of the maple with its skeletal winter form and leafy summer spread, or the country house with its own water tank and vegetable garden. These characteristics mean that both the maple and the rural family can withstand periods of shortage that would destroy a city or a rainforest.

Human health, water and food security are now under threat from the modest amount of climate change that has already occurred.

If humans keep doing things the same way for the first half of this century, I believe the collapse of civilisation due to climate change is inevitable.

Why have we done so little about global warming? We have known for some decades that the climate change we are creating for the twenty-first century is of a similar magnitude to that seen at the end of the last ice age, but that it is occurring thirty times faster. We have known that the Gulf Stream shut down on at least three occasions at the end of the last ice age, that sea levels rose by at least 100 metres, that agriculture was impossible before the Long Summer of 10,000 years ago.

What is the reason for our blindness? Is it an unwillingness to look such horror in the face and say, 'You are my creation.'

4

PEOPLE IN GREENHOUSES

TWENTY-TWO

THE STORY OF OZONE

A generation of children has now grown up knowing there is a hole in the ozone layer, which is why it's doubly important to wear sunscreen, sunglasses, and a hat in summer. This is a separate problem from increasing CO_2 but the story of ozone does show how we can co-operate internationally to solve huge environmental problems.

So what exactly is ozone, and why is it important? The gas that keeps your body alive consists of two atoms of oxygen joined together. But high in the stratosphere, ten to fifty kilometres above our heads, ultraviolet radiation occasionally forces an extra oxygen atom to join the duo. The result is a sky-blue gas known as ozone.

Ozone is unstable. It is constantly losing its additional atom, but new trios are forever being created by sunlight. This means a constant amount persists—about ten parts per million (one of every 100,000 molecules)—in an undamaged stratosphere. If all the planet's stratospheric ozone were brought down to sea level, it would form a layer just three millimetres thick.

Ozone is Earth's sunscreen. It shields us from around 95 per cent of the ultraviolet radiation that reaches Earth.

Without ozone's very high sun-protection factor, UV radiation would kill you fast, by tearing apart your DNA and breaking other chemical bonds within your cells.

The destruction of the ozone layer began long before anyone was aware of it. In 1928 industrial chemists invented CFCs (chlorofluorocarbons) and HFCs (hydrofluorocarbons). These inventions were found to be very useful for refrigeration, making styrofoam, as propellants in spraycans, and in air conditioning units. Their remarkable chemical stability (they do not react with other substances) made people confident that there would be few environmental side effects.

By 1975 spraycans alone were flinging 500,000 tonnes of the stuff into the atmosphere, and by 1985 global use of the main types of CFCs stood at 1.8 million tonnes. Their stability, however, was a key factor in the damage they caused, for they last a very long time in the atmosphere.

It is the chlorine in CFCs that is so destructive to ozone. Just a single atom can destroy 100,000 ozone molecules, and its destructive capacities are maximised at temperatures below –43°C. This is why the ozone hole first emerged over the South Pole, where the stratosphere is a frigid –62°C.

Researchers discovered that CFCs had raised chlorine levels in the stratosphere to *five* times their previous level.

The hole they punched in the ozone layer left people living south of 40° latitude exposed to a spectacular rise in the incidence of skin cancer. This includes people living in southern Chile and Argentina, in Tasmania and in the South Island of New Zealand.

At 53°S, Punta Arenas in Chile is the southernmost town on Earth. Since 1994 skin cancer rates there have soared by 66 per cent. Even nearer to the Equator—and closer to the great centres of human population—shifts in cancer rates are evident. In America, for example, the chance of getting melanoma was one in 250 just twenty-five years ago. Today it is one in 84, which is partly due to ozone depletion.

Ultraviolet radiation also causes damage to the immune system and the eyes. Researchers estimate that, for every 1 per cent decrease in ozone concentration, humans—and anything else with eyes—will experience a 0.5 per cent increase in cataracts. When people suffer cataracts the lenses of their eyes become opaque and they eventually go blind. As 20 per cent of cataracts are due to UV damage, the rate of blindness of this kind looks set to rise fast, especially among those who lack the means to protect themselves.

The impacts of UV increase will be felt throughout the ecosystem too. The microscopic single-celled plants that form the base of the ocean's food chain are severely affected by it, as are the larvae of many fish, from anchovy to mackerel. Indeed anything that spawns in the open is at risk.

Nor does agriculture escape its effects. The yield of crops such as peas and beans, for example, decreases 1 per cent for every extra per cent of UV radiation received.

Beginning in the 1970s, researchers began to warn of the calamity that was on the world's doorstep, though they did not yet have final proof of the relationship between CFCs and ozone destruction. Colour images of the ozone hole shown on television screens around the world convinced people that action needed to be taken, even if only as a precaution. Politicians were bombarded with letters requesting CFCs be banned.

DuPont was the main company responsible for their manufacture, and it and other companies launched a massive public relations campaign aimed at discrediting the link between their products and the problem.

Yet public concern remained high. Despite howls of protest from industry about cost, representatives of many countries met in Montreal in 1987 and signed the Protocol in which they agreed to phase out the offending chemicals. In that year final scientific proof of the link between CFCs and ozone depletion was announced.

Today we know just how important the Montreal Protocol was. Had it not been enacted, by 2050 the middle latitudes of the Northern Hemisphere (where most humans live) would have lost half of their UV protection, while equivalent latitudes in the Southern Hemisphere would have lost 70 per cent. As it was, by 2001, the Protocol had limited real damage to around a tenth of that.

Not all countries were initially bound by the Montreal Protocol. China continues to produce CFCs and may well go on polluting until 2010, when under the treaty it must cease. But production is limited because the newer substitute chemicals are so much better.

In 2004 the ozone hole over the Antarctic reduced by 20 per cent. Because the size of the hole changes from year to year, we cannot be certain that this decrease signals the end of the problem. Nevertheless, scientists are optimistic that in fifty years' time the ozone layer will be returned to its former strength.

The Montreal Protocol is our first ever victory over a global pollution problem.

Surely, with such a stunning all-round success behind them, the nations of Earth would have jumped at the chance to reduce global warming. And at first there was great enthusiasm for an international treaty to limit emissions of greenhouse gases. In 1997 the heads of many nations met in the Japanese city of Kyoto to hammer out such a treaty.

The meeting promised so much. So what happened?

TWENTY-THREE

THE ROAD TO KYOTO

The Kyoto Protocol is almost as famous as the hole in the ozone layer. It sets modest goals for reductions in CO_2 emissions of around 5 per cent. But four nations— USA, Australia, Monaco and Liechtenstein—have refused to ratify it, which would compel them to abide by its rules, and it has been bitterly contested. Why?

Supplying energy can be very profitable. In the developed world, energy use is growing at the rate of 2 per cent per annum or less. With such low rates of growth the only way for one sector (such as wind, gas or coal) to expand is to take from another sector's share. Kyoto will have a big influence on that outcome, and a furious struggle is ensuing between the potential winners and losers.

The Kyoto Protocol also marks a great divide between those who are certain it is essential to Earth's survival, and those who are fiercely opposed on economic and political grounds. Many in this latter group think Kyoto is economically flawed and politically unrealistic. Others believe that the entire climate change issue is hogwash.

Kyoto is in its infant stages, but despite the controversy

it's clear that it will influence all nations for decades to come.

The road to Kyoto began in 1985 with a scientific conference in Villach, Austria, which produced the first authoritative evaluation of the magnitude of climate change facing the world. In 1992, at the Rio Earth Summit, 155 nations signed the UN Framework Convention on Climate Change, which designated 2000 as the year by which signatory nations would reduce their emissions to 1990 levels. This target turned out to be wildly optimistic.

Following five years of negotiations, the signatories of the Framework Convention met in Kyoto, in Japan, in December 1997 and formed a new agreement about reducing emissions. The Kyoto Protocol had to be ratified by each country. It established two important things: greenhouse emissions targets for developed countries, and emissions trading of the six most important greenhouse gases, a trade which is valued at US$10 billion.

Emissions trading creates a new currency—a sort of 'carbon dollar'.

The targets allowed countries to create carbon budgets for themselves. By trading carbon—that is, paying for the right to pollute—industries can cost-effectively reduce their emissions. They can earn carbon credits by reducing their emissions, and sell these credits to more polluting industries. Companies which reduce emissions are rewarded,

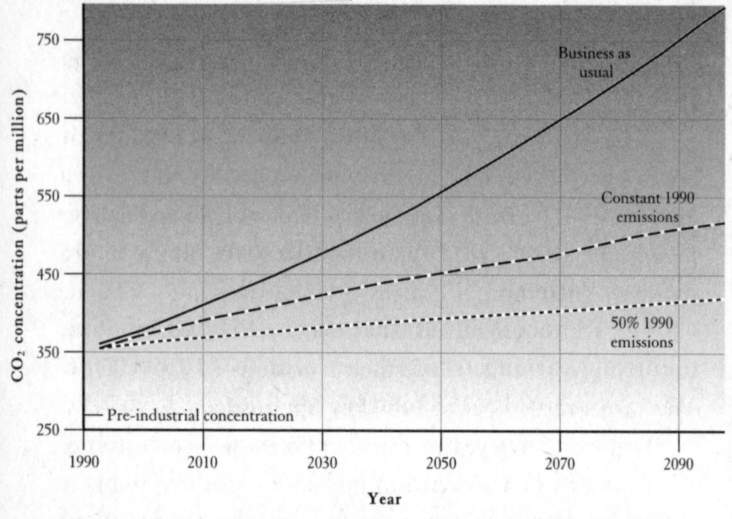

Large reductions in CO_2 are required to stabilise Earth's climate. The first phase of the Kyoto Protocol seeks to reduce emissions to 5 per cent less than 1990 levels, yet we would need to more than halve 1990-level emissions to stabilise CO_2 concentrations in the atmosphere.

and the products of those companies which continue to pollute become more expensive.

It sounds like a sensible scheme, yet it would take until late 2004, seven years after they agreed to it, for enough nations to ratify the treaty and propel it into life. The US and Australia refused to ratify it at all, even though they were part of the agreement in Kyoto.

Perhaps the most damning criticism of the Protocol is that it is a toothless tiger. Its targets amount to an average cut in CO_2 emissions between 1990 and 2012 of only

5.2 per cent. The momentum of climate change is now so great that this target is almost irrelevant.

If we are to stabilise our climate, Kyoto's target needs to be strengthened twelve times over: we need cuts of 70 per cent by 2050 to keep atmospheric CO_2 at double the pre-industrial level of 1800. This will be the challenge for future phases of the treaty.

The Protocol set different targets for participating countries, varying from 92 per cent to 110 per cent. Although they ratified, developing countries like China and India were exempted from having emissions targets during the first phase of the treaty (until 2012).

The matter becomes complex when the economies of the nations are taken into account. The eastern European states, for example, have suffered economic ruin since 1990 and are producing 25 per cent less CO_2 than they were at that time. Since their Kyoto targets are set at 8 per cent less than their 1990 levels, they now have valuable carbon credits to trade.

These credits, which do nothing to diminish climate change in the countries which buy them, are known as 'hot air'. They are a waste of dollars and a lost opportunity to reduce emissions. Many economists argue that the ex-communist countries should not be granted a steady stream of carbon dollars solely on the basis of poor economic performance.

For the first target period of the treaty up until the year

2012 the European Union has a carbon budget of 8 per cent less than it emitted in 1990 and the US 7 per cent less. Australia, on the other hand, has a budget of 8 per cent *greater* than it emitted back then. Only Iceland got a more generous target than that, with a 10 per cent increase. Was this a fair outcome, and how did it come about?

Australia has the highest per capita greenhouse emissions of any industrialised country—25 per cent higher than the US.

The Australian delegation to Kyoto argued that Australia's special circumstances—a heavy dependence on fossil fuels, special transport needs (being a large, sparsely populated continent) and an energy-intensive export sector—meant that the cost of meeting its Kyoto target was too high and therefore concessions were needed.

Ninety per cent of Australia's electricity is generated by burning coal. This is a matter of choice rather than necessity. Australia also has 28 per cent of the world's uranium, and the world's best source of geothermal power, where energy is derived from superheated water buried in rocks in the Earth's crust. Australia also has an abundance of high-quality wind and solar resources. Concerns about climate change have been voiced in Australia for over thirty years. For the nation to have remained so dependent on coal now seems like a poor economic decision. Should any country be rewarded for that?

The transport argument is also weak. Australia is large, but its population is highly urbanised, so 60 per cent of transport fuel is used in urban areas. And as for its energy-intensive exports, Australia is no more at risk here than Germany, Japan or the Netherlands—all strong Kyoto supporters.

No consensus had been reached by the scheduled end of negotiations in 1997, and the conference clock was stopped at midnight while delegates argued into the small hours. As the text was read for the final time, the leader of the Australian delegation, Senator Robert Hill, rose to his feet and raised a new issue: in Australia's case, land-clearing must be considered.

He argued that, by protecting forests, Australia was storing CO_2. As land-clearing had declined since the 1990 baseline year, this was similar to eastern European 'hot air'. It would mean Australia could comply with the Kyoto Protocol without having to do anything to minimise its carbon emissions. Faced with either agreeing to this request or seeing the meeting collapse, the delegates agreed to the concession.

Senator Hill was praised on his return to Australia, yet his country still refuses to ratify Kyoto, while claiming that it will reach its targets anyway! If you are confused by this, don't worry—so is the rest of the world. It's easy to get angry at such a self-interested, muddled approach to negotiations.

Australia's failure to ratify will be bad for trade. Japan—which purchases Australian coal—must now buy credits to offset emissions that result from burning that coal. But because Australia has not ratified the Protocol, no credits can be created there. Instead, the benefit of the credits will flow to a third country—perhaps New Zealand, which has ratified the treaty.

Those who endorse the new currency argue that carbon trading can dramatically lower the cost of meeting emission targets. And the use of emissions trading as a tool to reduce pollution has a good track record. The system was invented in the US in 1995 to deal with sulphur dioxide pollution from burning coal. It proved enormously successful and it has been adopted for a number of pollutants since.

This is how emissions trading works: a regulator imposes a permit requirement for the pollutant, and restricts the number of permits available. Permits are then given away on a proportional basis to polluters, or are auctioned off. Emitters who bear a high cost in reducing their pollution will then buy permits from those who can make the transition more easily. Benefits of the system include its transparency and ease of administration, the market-based price signal it sends, the opportunities for new jobs and products it creates, and the lowered cost of reducing pollutants.

For those who urge the abandonment of the Protocol or who criticise it, there are two questions: what do you propose to replace it with, and how do you propose to secure broad international agreement for your alternative? As yet, no answers to these questions have been put forward.

The Kyoto Protocol is the only international treaty in existence to combat climate change.

TWENTY-FOUR

COST, COST, COST

The governments of both the US and Australia say they refuse to ratify Kyoto because it would cost too much. A strong economy, they believe, offers the best insurance against all future shocks, and both are hesitant to do anything that might slow economic growth.

The emission decreases required to meet the first round of Kyoto targets to 2012 will be modest. This should assure us that compliance with Kyoto will not bankrupt our nations. It may even do our economies some good by directing investment into new infrastructure.

But to make a truly informed decision about Kyoto—or more radical proposals—we need to know the cost of doing nothing. Neither the US nor the Australian government has yet carried out this exercise.

The National Climatic Data Center in the US lists seventeen weather events that occurred between 1998 and 2002, which cost over a billion dollars apiece. They include droughts, floods, fire seasons, tropical storms, hailstorms, tornadoes, heatwaves, ice-storms and hurricanes. The most expensive, at a cost of US$10 billion, was the drought of

2002. All of these, of course, are nothing compared with the costs of hurricanes Rita and Katrina.

The costs of doing nothing about climate change are enormous.

Over the last four decades the insurance industry has suffered the burden of losses as a result of natural disasters. The impact of the 1998 El Niño offers a telling example. In the first eleven months of that year alone, weather-related losses totalled US$89 billion, while 32,000 people died and 300 million were made homeless. This was more than the total losses experienced in the entire decade of the 1980s.

Since the 1970s insurance losses have risen at an annual rate of around 10 per cent, reaching US$100 billion by 1999. Such a rate of increase implies that by about 2065, the damage bill resulting from climate change may equal the total value of everything that humanity produced in the course of a year.

Some time this century the day will arrive when human influences on the climate will overwhelm all natural factors. We will no longer be able to talk of climatic Acts of God, because any of us could have foreseen the consequences of what we are doing to our climate by continuing to pour CO_2 into it. Our legal system will have to decide who is to blame for human actions resulting from the new climate.

It's begun already. In 2004, the Inuit, who number 155,000 people, sought a ruling from the Inter-American

Commission on Human Rights about the damage global warming is doing to them.

Pretend, for a moment, that you are an Inuit teenager living in the Arctic. You are experiencing a rate of climate change twice that of the global average. Young men who drive the trucks carrying vital supplies to remote settlements along winter 'ice roads' are falling through the too-thin ice into lakes. In the winter of 2005–06 five died this way, prompting Inuit elders to say that climate change is now killing their young men. Your traditional food—seals, bear and caribou—are vanishing, and your land, in some instances, is disappearing under your feet. What would you do?

The Alaskan village of Shishmaref is becoming uninhabitable due to rising temperatures that are reducing sea ice and thawing permafrost, making the shoreline vulnerable to erosion. Hundreds of square metres of land and over a dozen houses have already been lost to the sea, and there are plans to relocate the whole town—at a cost of over US$100,000 per resident.

Shishmaref's plight is poignant. Its population is only 600 strong, but it has persisted at least 4000 years, and its inhabitants look set to become the first climate change refugees. Where they will go?

A favourable ruling from the Commission may enable Inuit either to sue the US government or US corporations. In either case it is likely that they will refer to the Universal

Declaration on Human Rights, which states that 'everyone has the right to a nationality' and that 'no one shall be arbitrarily deprived of his property', and to the United Nations Covenant on Civil and Political Rights, which states 'that in no case may a people be deprived of its own means of subsistence'.

Other inhabitants of lands immediately vulnerable to climate change are those of the five sovereign atoll countries in the Pacific. Atolls are rings of coral reef that surround a lagoon, and scattered around the reef crest are islands and islets, whose average height above sea level is a mere two metres. Kiribati, Marshall Islands, Tokelau and Tuvalu—which between them support around half a million people—consist only of atolls.

As a result of the destruction of the world's coral reefs, rising seas and intensifying weather events, it seems inevitable that these nations will be destroyed by climate change this century.

In the lead-up to the Kyoto summit, Australia insisted its Pacific Island neighbours drop their stance that the world should take 'firm measures' to combat climate change. 'Being small, we depend on them so much we had to give in,' said Tuvalu's Prime Minister Bikenibu Paeniu, following the South Pacific Congress in which Australia laid its demands on the table.

In what must be one of the most outrageous comments uttered in this context, the Australian government's chief economic adviser on climate change, Dr Brian Fisher, said that it would be 'more efficient' to evacuate small Pacific Island states than to require Australian industries to reduce their emissions of CO_2.

Any solution to the climate change crisis must be based upon principles of natural justice. And if democratic governments won't act voluntarily according to these principles, the courts may force them to do so. The principle of 'polluter pays' will become vital because the polluter should compensate the victim.

Prior to Kyoto all individuals possessed an unlimited right to pollute the atmosphere with greenhouse gases. The 162 nations who have ratified the Protocol now have an internationally recognised right to pollute within limits. Where does this leave the US and Australia who refused to ratify the treaty?

We still remain reluctant to tackle climate change. If scientists were predicting the imminent return of the ice age I'm certain our response would have been more robust.

'Global warming' creates an illusion of a comfortable, warm future. We are an essentially tropical species that has spread into all corners of our globe, and cold has long been our greatest enemy. From the beginning we have associated it with discomfort, illness and death, while warmth is the essence of everything good—love, comfort and life itself.

Our evolutionary response to the threat of cold is most clearly seen in the young. Children dredged from frozen ponds hours after slipping in have lived because, over millennia, their bodies have evolved defences against the ever-present threat of freezing to death. And, of course, parents, even in this modern age, do everything they can to protect their offspring from cold.

Our deep psychological resistance to thinking that 'warm' might be bad allows us to be deceived about the nature of climate change. This blind spot has left many people—even the well-educated—confused.

TWENTY-FIVE

PEOPLE IN GREENHOUSES SHOULDN'T TELL LIES

The opposition to reducing emissions of greenhouse gases is most intense in the US. The American energy sector is full of cashed-up businesses that use their influence to combat concern about climate change, to destroy emerging challengers, and to oppose moves towards greater energy efficiency.

In the 1970s the US was a world leader and innovator in energy conservation, photovoltaics (converting light to energy) and wind technology. Today it lags behind other countries in these areas. Over the past two decades some in the fossil-fuel industry have worked tirelessly to prevent the world from taking serious action to combat climate change.

The US coal producers have been centre-stage in this campaign. In the 90s Fred Palmer, now company vice-president at Peabody Energy, the world's largest coal producer, led a campaign that the Earth's atmosphere 'is deficient in carbon dioxide'. Producing more would bring in an age of eternal summer. Rather like the CEO of an arms manufacturer arguing that a nuclear war would be good for the planet, Peabody Energy wanted to create a world with atmospheric CO_2 of around 1000 parts per million.

Palmer's views were the basis for the propaganda video *The Greening of Planet Earth* which promoted the idea of 'fertilising' the world with CO_2 to boost crop yields by 30 to 60 per cent, thus bringing an end to world hunger. While such ludicrous claims could be laughed off by scientists, many people were misled.

On the other hand, some fossil-fuel companies are playing an active role in combating climate change. BP, for instance, has taken a clear-eyed view of climate change and has moved 'beyond petroleum', making a 20 per cent cut in its own CO_2 emissions, and a profit in doing so. BP has now become one of the world's largest producers of photovoltaic cells.

The British Prime Minister Tony Blair has a firm grasp of the science surrounding the issue. He has described global warming as 'a challenge so far-reaching in its impact and irreversible in its destructive power, that it alters radically human existence...There is no doubt that the time to act is now.'

By 2003 Britain's CO_2 emissions had fallen to 4 per cent below what they were in 1990. Significant milestones of this period include the establishment of the Carbon Trust (which helps business address energy use), an obligation by power suppliers to provide 15.4 per cent of their energy from renewable sources, and significant investments in developing wave and tidal power. Britain is also considering expanding its nuclear power capacity.

These debates about how to transfer from fossil fuels to renewable sources of energy will only grow more intense.

Can we find solutions to the problem of global warming while continuing to use fossil fuels?

The coal industry is promoting the idea of pumping CO_2 underground in order to take it out of the atmosphere. The process, known as geosequestration—it means hiding in the earth—is simple in its approach: the industry would bury the carbon that it had dug up.

Oil and gas companies have been pumping CO_2 underground for years. A good example is the Norwegian Sleipner oilfield in the North Sea where about a million tonnes of CO_2 is pumped underground each year. The Norwegian government has placed a US$40 per tonne tax on CO_2 emissions. This provides the incentive at Sleipner to separate out the CO_2 that comes up with the oil and pump it back into the rocks.

At a few other wells around the world, the CO_2 is pumped back into the oil reserve, helping to maintain head pressure, which assists with the recovery of oil and gas, making the entire operation more profitable. The companies claim 'most' of the CO_2 stays underground. Applying this model to the coal industry, however, is not straightforward.

The problems for coal commence at the smokestack. The stream of CO_2 emitted there is relatively dilute, making its capture unrealistic. The coal industry is promoting a new process known as coal gasification, which creates a more concentrated stream of CO_2 for capture and burial. These plants are not cheap to run: around one quarter of the energy they produce is consumed just in keeping them operating. Building them on a commercial scale will be expensive and it will take decades for them to make a big contribution to power production.

Let's assume that some plants are built and the CO_2 they emit is captured. For every tonne of anthracite burned, around 3.7 tonnes of CO_2 is generated, all of which must be stored. The rocks that produce coal are not often useful for storing CO_2, so the gas would have to be transported away from the power stations. In the case of Australia's Hunter Valley coal mines, for example, it would need to be carried over Australia's Great Dividing Range and hundreds of kilometres to the west to a suitable site.

Once the CO_2 arrives at its destination it must be compressed into a liquid so it can be injected into the ground—a step that typically consumes 20 per cent of the energy yielded by burning coal in the first place. Then a kilometre-deep hole must be drilled and the CO_2 injected. From that day on, the geological formation must be closely monitored. If the gas were ever to escape, it has the potential to kill.

Miners used to call concentrated CO_2 'choke-damp', an appropriate name as it instantaneously smothers its victims.

The largest recent disaster caused by CO_2 occurred in 1986, in Cameroon, central Africa. A volcanic crater-lake known as Nyos belched bubbles of CO_2 into the still night air and the gas settled around the lake's shore. It killed 1800 people and countless thousands of animals, both wild and domesticated.

No one is suggesting we bury CO_2 in volcanic regions like Nyos, so the CO_2 dumps created by industry are unlikely to cause a similar disaster. Still, Earth's crust is not a purpose-built vessel for holding CO_2, and the storage must last thousands of years. The risk of a leak must be taken seriously.

The amount of CO_2 we would need to bury is mind-boggling. We can use a country like Australia, with its comparatively small population, as an example. Imagine a pile of 200-litre drums, ten kilometres long and five kilometres across, stacked ten drums high. That would be more than 1.3 billion drums, the number required to hold the CO_2 that pours out of Australia's twenty-four coal power stations, which provide power to 20 million people *every day*. Even when compressed to liquid form, that daily output would take up a third of a cubic kilometre, and Australia accounts for less than 2 per cent of global emissions!

Imagine injecting 20 cubic kilometres of liquid CO_2 into the Earth's crust every day of the year for the next century or two.

If we were to try to bury all the emissions from coal, the world would very quickly run out of A-grade reservoirs near power stations. There are enough fossil fuel reserves on planet Earth to create 5000 billion tonnes of CO_2. How could Earth tuck that away without suffering fatal indigestion?

The best-case scenario for geosequestration is that it will play a small role (at most perhaps 10 per cent by 2050) in the world's energy future.

There are other forms of sequestration—of hiding carbon—which are vital for the future of the planet, and which carry no risk. Earth's vegetation and soils are reservoirs for huge volumes of carbon, and are critical elements in the carbon cycle. Today the world is mostly deforested and its soils exhausted, but soil carbon can be enhanced by following sustainable agricultural and animal farming practices.

This increases the vegetable mould (mostly carbon) in the soil. Lots of carbon—around 1180 gigatonnes—is currently stored this way; more than twice as much as is stored in living vegetation (493 gigatonnes). There is real hope for progress here, in initiatives that include everything from organic market gardening to sustainable rangelands management.

Can we store carbon in forests and long-lived forest products? This involves either planting forests, or preserving them. The Costa Rican government saved half a million hectares of tropical rainforest from logging. This brought it carbon credits equivalent to the amount of CO_2 that would have entered the atmosphere if the forests had been disturbed.

Another example is BP's initiative to fund the planting of 25,000 hectares of pine trees in Western Australia to offset emissions from its refinery near Perth.

Forestry plantations are destined to be cut and used, but they can be a good short-term store for carbon because the furniture and housing they produce are long-lived, and because the roots of the felled trees (along with their carbon) stay in the ground.

The carbon in coal has been safely locked away for hundreds of millions of years, and will remain there for millions more if we refuse to dig it up.

Carbon locked away in forests or the soil is unlikely to remain out of circulation for more than a few centuries. By trading coal storage for tree storage of carbon, we are exchanging a long-term guarantee for a quick fix.

Scientists continue to work on the problem of safe, secure storage for carbon, and perhaps a solution will eventuate. Meanwhile, the competition from less carbon-dense fuels is looking simpler and cheaper by the day.

TWENTY-SIX

LAST STEPS ON THE STAIRWAY TO HEAVEN?

For many people the solution to the climate change problem is like climbing an imaginary staircase of fuels. Each step contains an ever-diminishing amount of carbon.

Yesterday, the argument goes, it was coal, today it's oil and tomorrow it will be natural gas. Heaven will be reached when the global economy makes the transition to hydrogen—a fuel that contains no carbon at all.

Technological advances, high oil prices, a looming lack of oil and the demand for a cleaner fuel to replace coal have all combined to change the economics of gas. Today it is big business. The most important advance involves the refrigeration of gas so that it becomes a supercooled liquid, which permits cost-effective transport, in purpose-built ships, over large distances. There is an international trade in shipping, and big companies are investing the billions required for pipelines, so gas appears to be the fuel of choice for the twenty-first century.

Although gas is a more expensive fuel than coal, it has

many benefits that make it ideal for producing electricity. Gas-fired power plants cost half as much to build as coal-fired models, and they come in a variety of sizes. Instead of having one massive, distant generator of electricity, as with coal, a series of small gas-fired generators can be dotted about, saving on transmission losses. They can also be fired up and shut down quickly, which makes them an ideal complement for intermittent sources of energy generation such as wind and solar.

Over 90 per cent of new power generation in the US today is gas-fired, and around the world it is fast becoming the favoured fuel. But gas has its problems, including safety issues and the possibility of terrorist attacks on large plants or pipelines. And because methane is a powerful greenhouse gas, we must be aware of its potential to leak. The old iron pipes used to send gas throughout cities are often leaky.

If all the coal-fired power stations on Earth were replaced with gas-fired ones, global carbon emissions would be cut by only 30 per cent. If we were to stall on this step of the energy staircase, we would still face massive climate change.

In this scenario, a transition to hydrogen is imperative. How likely is it?

Ever since the phrase 'hydrogen economy' was coined, hydrogen has presented itself to many people as the silver-bullet solution to the world's global warming ills. But there's a lot of devil in the detail.

It's important to understand that hydrogen is an energy carrier—like a battery. The energy it holds has to be created from another source, and if that source is a fossil fuel, then CO_2 emissions will be created in the process.

The power source of the hydrogen economy is the hydrogen fuel cell, which is basically a box with no moving parts that takes in hydrogen and oxygen, and puts out water and electricity.

The most promising cells for the stationary production of electricity are known as molten carbonate fuel cells, which operate at a temperature of around 650°C. They are highly efficient but they take a long time to reach working temperature. They are also very large—a 250-kilowatt model is the size of a railway carriage—making them unsuitable for use in vehicles.

How could we use hydrogen as a transport fuel? This requires smaller fuel cells that work at lower temperatures. A number of car manufacturers, including Ford and BMW, are planning to introduce hydrogen-fuelled cars to the marketplace. And the Bush administration plans to invest US$1.7 billion to build the hydrogen-powered FreedomCAR.

In the hydrogen economy we could refuel our vehicles from hydrogen pumps at fuel stations. The hydrogen could be produced at a remote central point and distributed to fuelling stations; but it's here that the difficulties become evident.

The ideal way to transport it is in tanker-trucks carrying liquefied hydrogen. But liquefication occurs at –253°C, so refrigerating the gas to achieve this is an economic nightmare. Using hydrogen energy to liquefy a kilogram of hydrogen consumes 40 per cent of the value of the fuel. Using the power grid to do so takes 12–15 kilowatt hours of electricity, and this would release almost ten kilograms of CO_2 into the atmosphere. Around 3.5 litres of petrol holds the equivalent energy of one kilogram of hydrogen. Burning it releases around the same amount of CO_2 as using the grid to liquefy the hydrogen.

So the climate change consequences of using liquefied hydrogen could be as bad as driving a standard car.

One solution may be to pressurise the hydrogen only partially, which reduces the fuel value consumed to 15 per cent, and the canisters used for transport can be less specialised. But even using improved canisters, a 40-tonne (40,000 kilogram) truck could deliver only 400 kilograms of compressed hydrogen, so it would take fifteen such trucks to deliver the same fuel energy value as is now delivered by a 26-tonne petrol tanker. And if these 40-tonne trucks carried the hydrogen 500 kilometres, the energy cost of the transport would consume around 40 per cent of the fuel carried.

Further problems arise when you store the fuel in your car. You would need a special high-pressure fuel tank ten times the size of a petrol tank. Around 4 per cent of its fuel is likely to be lost to boil-off every day. The main tank of the space shuttle, for example, takes 100,000 litres of hydrogen, but NASA needs to deliver an extra 45,000 litres at each re-fuelling just to account for the evaporation rate.

Pipelines are another option for transporting hydrogen, but they are expensive and must be of high integrity, because hydrogen leaks so easily. Even if the gas pipeline network could be reconfigured to transport hydrogen, the cost of providing a network running from central producing units to the world's fuelling stations would be astronomical.

Perhaps hydrogen could be produced from natural gas at the fuelling station. This would do away with the difficulties of transporting it, but would produce 50 per cent more CO_2 than using the gas to fuel the vehicle in the first place.

Hydrogen could also theoretically be generated at home using power from the electricity grid, but it would be far too expensive. The electricity in the grid in places such as the US is largely derived from burning fossil fuels, so home generation of hydrogen under current circumstances would result in a massive increase in CO_2 emissions.

And there is another danger with home-brewing hydrogen. The gas is odourless, leak-prone, highly combustible and it burns with an invisible flame.

Firemen are trained to use straw brooms to detect a hydrogen fire: when the straw bursts into flames you have found your fire.

Let's imagine for a moment, however, that all of the delivery problems relating to hydrogen are overcome, and you find yourself at the wheel of your new hydrogen-powered four-wheel-drive. Your fuel tank is large and spherical, because at room temperature hydrogen takes up around three thousand times as much space as petrol. Now consider that the static electricity generated by sliding over a car seat, or even an electrical storm 1.6 kilometres away, carries a sufficient charge to ignite your hydrogen fuel. A hydrogen car accident is too horrible to think about.

Even if hydrogen is made safe to use, we are still left with a colossal CO_2 pollution issue, which was exactly the opposite of what we set out to do.

The only way that the hydrogen economy can help combat climate change is if the electricity grid is powered entirely from carbon-free sources.

5

THE SOLUTION

TWENTY-SEVEN

BRIGHT AS SUNLIGHT, LIGHT AS WIND

In our war on climate change we must decide whether to focus our efforts on transport or the electricity grid. If we decarbonise the grid first, we can use the renewable power we generate to decarbonise transport.

Two researchers from Princeton University in the US investigated whether the world possesses the technologies required to run an electricity network like the one we currently enjoy, while at the same time making deep cuts in CO_2 emissions.

They identified fifteen basic kinds of technologies, ranging from sequestration to wind, solar and nuclear power, which are developed now and could play a vital role in controlling the world's carbon emissions for at least the next fifty years.

There are many examples of governments and corporations around the world that have slashed emissions (by over 70 per cent in the case of some British local councils) while at the same time experiencing strong economic growth.

The technologies fall into two lots: those that provide power intermittently; and those that provide a continuous output of power.

Of all the sources of intermittent power, the most mature and economically competitive is wind. And Denmark is the home of the modern wind industry.

At the time the Danes decided to back wind power, the cost of electricity produced this way was many times greater than that produced by fossil fuels. The Danish government, however, could see the potential of wind power and supported the industry until costs came down.

Denmark leads the world in both wind power production and the building of turbines.

Wind now supplies 21 per cent of Denmark's electricity. Around 85 per cent is owned by individuals or wind co-operatives. Power lies literally in the hands of the people.

In several countries wind power is already cheaper than electricity generated from fossil fuel, which helps account for the industry's phenomenal growth rate of 22 per cent per annum. It has been estimated that wind power could provide 20 per cent of the energy needs of the United States. Over the next few years the unit price of wind energy is expected to drop a further 20–30 per cent, which will make it even more cost effective.

Wind power is widely perceived as having a major disadvantage—the wind doesn't always blow, which means

Wind farms are increasingly contributing to our global electricity grid. Between 2005–08, China will install at least 3000 megawatts of wind power. That's enough to power a small city.

that it is unreliable. It's true that the wind does not blow at the same place with consistent strength, but if you take a regional approach it is fairly certain that the wind will be blowing somewhere. As this suggests, there is a lot of redundancy in wind generation, for often there will be several turbines lying idle for each one working at full strength.

In the UK the average turbine generates at only 28 per cent of its capacity over the course of a year. But all forms of power generation have some degree of redundancy. In the

UK nuclear power works at around 76 per cent, gas turbines 60 per cent, and coal 50 per cent of the time. This disadvantage in wind is somewhat offset by its reliability: wind turbines break down less often and are cheaper to maintain than coal-fired power plants.

Wind power, unfortunately, has received bad press, including allegations that wind turbines kill birds, and are noisy and unsightly. The truth is, any tall structure represents a potential hazard to birds, and early wind towers did increase that risk—they had a latticework design, allowing birds to nest in them. But they have now been replaced by smooth-sided models.

All risks need to be measured against each other. Cats kill far more birds in the US than do wind farms. And if we continue to burn coal, how many birds will die as a consequence of climate change?

As for noise pollution, you can have a conversation at the base of a tower without having to raise your voice, and new models reduce the sound even further. And in terms of their ugliness, beauty is surely in the eye of the beholder. What is more unsightly—a wind farm or a coal mine and power plant? Besides, none of these issues should be allowed to decide the fate of our planet.

Let's turn to the Sun and three technologies that directly exploit its power. These are solar hot water systems, solar-thermal devices and photovoltaic cells. Solar hot water is the simplest and, in many circumstances, the best way to make large, easy savings in most household power bills. In the Southern Hemisphere solar hot water systems sit on a north-facing roof (in the Northern Hemisphere south-facing), and trap the Sun's rays that are then used to heat water. They require no maintenance and, to ensure that hot water is available whenever you need it, they include a gas or electric booster.

Solar-thermal power stations produce large amounts of electricity—far more than one household could ever use—and they work by concentrating the Sun's rays onto small, highly efficient solar collectors. Their name comes from the fact that they produce both electricity and heat. There are many designs in the marketplace at present, and they are rapidly becoming more affordable. In future, solar-thermal power plants can be expected to compete with wind for a slice of the grid. Wind and solar-thermal are perfect partners—if the wind isn't blowing, there's a good chance that the Sun will be shining.

Finally there is the technology that most people recognise as true 'solar' power—photovoltaic cells. Generating your own electricity with photovoltaics is liberating—once you've bought your equipment you don't have to depend on the big power companies. It is also simple and maintenance-free, unless you are not connected to the grid and need a battery bank.

Photovoltaic cells use the sunlight that falls on them to generate electricity. The average home requires around 1.4

kilowatts (1400 watts) of power to run, and the average size of panels is 80 or 160 watts. Ten of the larger size should do the job, though it is amazing how much more power conscious you become (and thus how much power you save) when you are generating your own.

Photovoltaics operate best in summer, when that extra power for air conditioning is needed. This allows the owner of a photovoltaic system to make money: in Japan you can sell excess power to the grid for as much as US$50 per month, and similar schemes exist in fifteen other countries. The cost of photovoltaics is falling so rapidly that electricity generated by this means is expected to be cost effective as early as 2010.

There are, of course, many kinds of renewable power generation I haven't discussed, including solar chimneys and tidal and wave power, and in certain locations all of these options are now, or soon will be, producing power.

If renewable energy offers one lesson, it is that there is no single solution for taking carbon out of the power grid: instead it presents us with many technologies to choose among. These technologies exist now. We can choose those that suit us best to cut our carbon emissions by 70 per cent by 2050.

TWENTY-EIGHT

NUCLEAR?

It's sometimes said that the Sun is nuclear energy at a safe distance, but as we know, even though it's far away, it can still burn us. In this era of climate crisis, however, the role of Earth-based nuclear power is changing. What was until recently a dying technology may yet make a comeback.

The revival began in May 2004, when environmental organisations around the world were shocked to hear the originator of the Gaia hypothesis, James Lovelock, deliver a heartfelt plea for a massive expansion in the world's nuclear energy programs as a way to combat climate change. He compared our present situation with that of the world in 1938—on the brink of war and nobody knowing what to do. Organisations such as Greenpeace and Friends of the Earth immediately rejected his call.

Yet Lovelock has a point, for all power grids need reliable 'baseload' generation, and there remains a big question mark over the capacity of renewable technologies to provide it. France supplies nearly 80 per cent of its power from nuclear sources, while Sweden provides half and the UK one quarter.

Nuclear power already provides 18 per cent of the world's electricity, with no CO_2 emissions. Its defenders argue that it could supply far more.

Nuclear power plants are nothing more than complicated and potentially hazardous machines for boiling water, which creates the steam used to drive turbines.

As with coal, conventional nuclear power stations are very large—around 1700 megawatts—and with a starting price of around US$2 billion apiece they are expensive to build. The cost of the power they generate, however, is at present competitive with that generated by wind. But they take ten years to approve and five to build. With a fifteen-year period before any power is generated, and even longer before any return on the investment is seen, nuclear power is not for the impatient. No new reactors have been built for twenty years in either the US or the UK.

Three factors loom large in the mind of the public whenever nuclear power is mentioned—safety, disposal of waste, and bombs. The 1986 Chernobyl disaster in Ukraine was a catastrophe whose consequences, two decades after the accident, just keep growing. In Belarus, which received 70 per cent of the fallout, only 1 per cent of the country is *free* from contamination, and 25 per cent of its farmland has been put permanently out of production.

In the US and Europe, safer reactor types predominate but, as the 1979 Three Mile Island incident in Pennsylvania shows, no one is immune to accident, or to sabotage. With several nuclear reactors in the US located near large cities, there are real concerns about a possible terrorist attack.

The management of radioactive waste is another difficulty. And the problem of what to do with old and obsolete nuclear power plants is almost as hard: the US has 103 nuclear plants that were originally licensed to operate for thirty years, but are now required to grind on for double that time. This ageing fleet must be giving the industry headaches, especially as no reactor has ever yet been successfully dismantled, perhaps because the cost is estimated to be around US$500 million a pop. All of these disadvantages, however, need to be balanced against the alternatives. Each year, for example, coal mining and coal-fired power plants kill far more people, in mine accidents and through lung disease, than uranium mining and nuclear power plants.

Most new nuclear power plants are being built in the developing world. China will commission two new nuclear power stations per year for the next twenty years, which from a global perspective is highly desirable, for 80 per cent of China's power now comes from coal. Indeed China will soon open the world's first pebble bed reactor, which is a very safe and efficient kind of small (300 megawatt) nuclear power plant.

India, Russia, Japan and Canada also have reactors under construction, while approvals are in place for thirty-seven more in Brazil, Iran, India, Pakistan, South Korea, Finland and Japan.

Providing the uranium necessary to fuel these reactors will be a challenge, for global reserves are not large, and at the moment around a quarter of the demand is being met by reprocessing redundant nuclear weapons. This brings us to the issue of nuclear weapons getting into the wrong hands. Anyone who possesses enriched uranium has the ability to create a bomb. As reactors proliferate and alliances shift, it is more likely that such weapons will be available to those who want them. Only good regulation and a willingness to support international treaties and the agencies (such as the International Atomic Energy Agency) that police the regulations can minimise these dangers.

What role might nuclear power play in preventing a climate change disaster? China and India are likely to pursue the nuclear option with vigour, for there is currently no inexpensive, large-scale alternative available to them. Both nations already have nuclear weapons programs, so the relative risk of proliferation is not great. In the developed world, though, any major expansion of nuclear power will depend upon the viability of new, safer reactors.

Geothermal is another option for the continuous production of power.

Geothermal describes the reservoirs of heat lying between our feet and our planet's molten mantle. As the inventor of the commercial alternating current, Nikola Tesla, first observed, there's clearly a lot of heat under our feet, but geothermal technologies provide a mere 10,000 megawatts of power worldwide. Why? It turns out that we have been looking for heat in the wrong places. Previously geothermal power has come from volcanic regions, where aquifers (underground reservoirs) flowing through the hot rocks provide superheated water and steam. It seems sensible to seek geothermal power in such places, but consider the geology.

Lava volcanoes only exist where the Earth's crust is being torn apart, allowing the magma below to come to the surface. Iceland, which formed from the ocean floor when Europe and North America drifted away from each other, is an excellent example of this. There is plenty of heat in such places, but the biggest problem is the aquifers. Although many flow freely when first tapped, they quickly dwindle, leaving the power plant without a means of transferring the rock's heat to its generators. In the 1980s operators began to pump water back into the ground in the hope that it would be reheated and could be reused. Quite often the water just vanished into vertical faults and was never seen again.

In Switzerland and Australia, companies are finding commercially useable heat in the most unlikely places. When oil and gas companies prospected in the deserts of northern South Australia, nearly four kilometres below the surface, they discovered a body of granite heated to around

250°C—the hottest near-surface, non-volcanic rock ever discovered.

What really excited the geologists was that the granite was in a region where the Earth's crust was being compressed. This led to horizontal, rather than vertical fracturing of the rock. Even better, the rocks are bathed in superheated water under great pressure, and the horizontal fracturing meant that it could be readily recycled.

This one rock body in South Australia is estimated to contain enough heat to supply all of Australia's power needs for seventy-five years, at a cost equivalent to that of brown coal, without the CO_2 emissions. So vast is the resource that distance to market is no object, for electricity can be pumped down the power line in such volume as to overcome any transmission losses.

With trial power plants scheduled for construction, the enormous potential of geothermal power is about to be tested. Geologists around the world are scrambling to prospect for similar deposits, as the extent of the resource is hardly known.

While it appears to be an exciting breakthrough, we must remember that so far very little electricity has been provided by this form of geothermal heat. It will probably be decades before this technology is contributing significantly to the world power grid.

The power technologies I have discussed place humanity at a great crossroad. What might life be like if we choose one over the other? In the hydrogen and nuclear economies the production of power is likely to be centralised, which would mean the survival of the big power corporations.

Pursuing wind and solar technologies opens the possibility that people will generate most of their own power, transport fuel and even water (by condensing it from the air).

Decarbonising the grid may literally decentralise power and empower individuals.

TWENTY-NINE

HYBRIDS, MINICATS AND CONTRAILS

So how do we go about decarbonising our transport systems? Transport, after all, accounts for around a third of global CO_2 emissions.

Among those pursuing renewables, the Brazilians have the lead. Much of their vehicle fleet runs on ethanol, a fermented alohol produced from sugar cane—which grows extremely well in Brazil. One third of new cars sold in Brazil last year could run on either ethanol or petrol, allowing the consumer to choose the cheapest. And they cost no more than standard models. In the US ethanol is largely derived from corn, but the amount of fossil fuel put into growing the crop means that the use of corn-derived ethanol in transportation provides little in the way of carbon savings.

If a highly efficient source of ethanol—perhaps switchgrass—can be cultivated, the crop would have to make up 20 per cent of all productivity on land to power the world's cars, ships and aircraft. Humans are already consuming more of the planet's resources than is sustainable, so providing this extra biological productivity will be hard, and will depend on the development of more sustainable agriculture.

Despite such obstacles, technological advances in transport are so rapid that ways forward can be glimpsed, and nowhere is that so clear as in Japan.

While American companies like Ford have been investing in hydrogen, Toyota and Honda have been hiring engineers to design more efficient cars. They have discovered a revolutionary new technology which halves fuel consumption and opens the way to astonishing future developments. Known as hybrid fuel vehicles, these new automobiles pair a petrol-driven engine with a revolutionary electric motor.

Travelling in the Toyota Prius is unnerving at first; it can be so quiet you think that the engine has stopped. Instead, when slowing or stopped in traffic, the 1.5-litre petrol engine shuts down and the silent electric motor takes over. This is powered by energy generated in part from braking—energy wasted in an ordinary vehicle. The Prius has taken the market by storm and, with a tank that needs refuelling every thousand kilometres, it's the least carbon-costly automobile of its size available, or likely to be available for the next decade or two.

Relative to the Toyota Landcruiser (or other four-wheel-drives so popular in the US and Australia today) the Prius cuts fuel use and CO_2 emissions by around 70 per cent. That is the amount scientists consider is required for the world economy by 2050 in order to stabilise climate change.

If you wish to make a real contribution to combating climate change, don't wait for the hydrogen economy—get your family to buy a hybrid fuel car.

If the grid were to be decarbonised, many other transport options become attractive. Electric cars have been on the market for years, and France already has a fleet of 10,000 such vehicles. But even more exciting technologies are coming out of Europe, including the experimental compressed-air car.

Imagine what a compressed-air car might mean for a family living in Denmark. They may well own a share in a wind generator, which is used to power their home, and could be used to compress air for their transport fuel as well. Contrast this with the average American family who, even if nuclear and hydrogen options become available, will continue to purchase their electricity and transport fuels from large corporations. By combating climate change we can save our planet *and* open the way for a very different future as well.

What about other growing transport sectors such as shipping and air travel? One of the foulest pollutants on Earth is the fuel oil that powers shipping. Over the past few years the volume of international shipping has grown by 50 per cent, and cargo ships have become a leading source of air pollution. These vessels are powered by the leftovers from the production of other fuels. It is so thick and full of contaminants that it must be heated before it will even flow through a ship's pipes.

Satellite surveillance reveals that many of the world's shipping lanes are blanketed in semi-permanent clouds that pour from ships' smokestacks. Solving this problem is potentially easy. Until little more than a century ago, maritime transport was wind-powered. Using modern wind and solar technologies and energy-efficient engines, sea cargo could, by the middle of this century, once again be travelling carbon-free.

Air travel requires large amounts of high-density fuel that at present only fossil fuels provide. In 1992 air travel was the source of 2 per cent of CO_2 emissions but that is growing rapidly. And in the US, where air traffic already accounts for 10 per cent of fuel use, the number of passengers transported is expected to double between 1997 and 2017, making air transport the fastest growing source of CO_2 and nitrous oxide emissions in the country. Across the Atlantic, by 2030, a quarter of the UK's CO_2 emissions may come from air travel.

The chemicals that comprise aircraft emissions work in somewhat opposite ways. Because most modern jets cruise near the troposphere, the water vapour, nitrous oxide and sulphur dioxides they emit have particular impacts. The nitrous oxide emitted by aircraft may enhance ozone in the troposphere and lower stratosphere, yet deplete it further in the upper stratosphere. Sulphur dioxide will have a cooling effect.

Water vapour, which can be observed as aircraft contrails, may be hugely important. Under certain conditions contrails give rise to cirrus clouds. Cirrus clouds cover around 30 per cent of the planet. It could be that aircraft contribute as much as 1 per cent to cirrus cloud cover, which may have a significant warming impact on climate.

If aircraft were to fly lower, cirrus cloud formation could be cut in half and CO_2 emissions lowered by 4 per cent, while average flight times over Europe would vary by less than a minute.

While Europeans and Japanese may be able to switch from jets to fast trains for travel, for Australians, Canadians and Americans, there are no realistic alternatives. Aircraft will have to use fossil fuels for the foreseeable future. Without a return to the more leisurely days of travel by zeppelin, flying will remain a source of CO_2 emissions long after other sectors have transformed to a carbon-free economy.

THIRTY

OVER TO YOU

If everyone takes action to rid atmospheric carbon emissions from their lives, I believe we can stabilise and then save the Arctic and Antarctic. We could save around four out of every five species currently under threat, limit the extent of extreme weather events and reduce, almost to zero, the possibility of any of the three great disasters occurring this century, especially the collapse of the Gulf Stream and the destruction of the Amazon.

But for that to happen, everyone needs to act on climate change now: the delay of even a decade is far too much. Some big things are beyond the control of any single individual. We shouldn't, for instance, build or expand any old-fashioned coal-fired power plants. Indeed we must begin shutting them down. Those decisions will be made by governments, but government is much more likely to make the right decision if people demand it.

Whether you are old enough to vote or not, you can make politicians aware of your views. And if you've taken action in your own life to reduce emissions, you can ask others what they personally are doing to reduce theirs.

This is the single most important thing I want to say: there is no need to wait for government to act. You can do it yourself. The technology exists to reduce the carbon emissions of almost every household on the planet.

You can in a few months easily attain the 70 per cent reduction in emissions required to stabilise the Earth's climate. All it takes are a few changes to your life and your family's life, none of which requires serious sacrifices.

Understanding how you use electricity is the most powerful tool in your armoury, for that allows you to make effective decisions about reducing your personal emissions of CO_2.

Have you ever looked at your family's electricity bill? If not, ask to see it and read it carefully. Is there a green power option, where the electricity company guarantees that the power that flows into your home will come from renewable sources like wind or solar or hydro? The green power option can cost as little as a dollar per week, yet is highly effective in reducing emissions.

If your electricity company does not offer a suitable green option, suggest your family dump them and call a competitor. Changing your power supplier is usually a matter of a single phone call, involving no interruption of supply or inconvenience in billing.

It is possible then, in switching to green power, for your family to reduce its household emissions to zero. All as the result of a single phone call.

What about hot water? In the developed world, roughly one third of CO_2 emissions result from domestic power, and one third of a typical domestic power bill is spent on heating water. This is crazy, since the Sun will heat your water for free if you have the right device.

Your family will need to make an initial outlay, but such are the benefits that it is well worth taking out a loan to do so, for in sunny climates like Australia or California or southern Europe the payback period is around two or three years. The solar heating devices usually carry a ten-year guarantee, so you will get at least seven to eight years of free hot water. Even in cloudy regions such as Germany and Britain you will still get several years' worth of hot water for free.

Then there's air conditioning, heating and refrigeration which chew up the most power. Make sure your family researches the most energy-efficient model available. It may be cheaper to install insulation rather than buying and running a larger heater or cooler.

Suggest that your family reads its power bill together and set a target for reduction. If you meet the target the savings could go towards a family holiday.

I became so outraged at the irresponsibility of the coal burners that I decided to generate my own electricity, which has proved to be one of the most satisfying things I've ever done. For the average family, solar panels are the best way to do this. Twelve 80-watt panels is the number I granted myself, and the amount of power this generates in Australia is sufficient to run the house.

To survive on this amount, our family is vigilant about energy use, and we cook with gas. And I'm fitter than before because I use hand tools rather than the electrical variety to make and fix things. Solar panels have a twenty-five-year guarantee (and often last for up to forty years). I'll be enjoying the free power they provide well into retirement.

The town of Schoenau in Germany offers a different example of direct action. Some of its families were so alarmed by the Chernobyl disaster that they decided to do something to reduce their dependence on nuclear power. It started with a group of ten parents who gave prizes for energy savings. This proved so successful that it soon bloomed into a citizens' group determined to take control of the town's power supply from KWR, the monopoly that supplied them.

They put together their own study, then raised the money to build their own green power scheme. Eventually they raised enough to purchase the power supply, grid and all, from KWR. Today the town not only runs its own power supply but a successful consulting business which advises on how to 'green' the grid right across the country. Each year Schoenau's power supply becomes greener, and even the big power users, such as a plastics recycling factory situated in the town, are happy with the result.

Wouldn't it be wonderful if you could start a movement like this in your town or neighbourhood?

It is not possible right now for most of us to do away with burning fossil fuels for transport, but we can greatly reduce their use. Walking or riding a bicycle wherever possible—to school or work or to the shops—is highly effective, as is taking public transport. If your family trades in its four-wheel-drive or SUV for a medium-sized hybrid fuel car you will have cut your personal transport emissions by 70 per cent overnight.

For those who cannot or do not wish to drive a hybrid, a good rule is for your family to buy the smallest vehicle capable of doing the job it most often requires. You can always rent for the rare occasions you need something larger. A few years from now, if you have invested in solar power, you should be able to purchase an electric or compressed-air vehicle. Then, your family can forget all of those power and petrol bills.

Despite the way it often feels, students and employees wield considerable influence in schools and workplaces. To become more greenhouse aware, ask for an energy audit or review to be done. That's how you can be sure your use of energy is as efficient as possible.

And remember, if you can cut your personal emissions by 70 per cent, so can schools, companies, farms and many other organisations.

Society desperately needs advocates—people who understand the issues and will act and serve as witnesses to what can and should be done. By taking such public actions you will be achieving results that extend way beyond their local impact.

As you read through this list of actions to combat climate change, you might doubt that such steps can make such a big difference. But if enough of us buy green power, solar panels, solar hot water systems and hybrid vehicles, the cost of these items will fall. This will encourage the sale of yet more panels and wind generators, and soon the bulk of domestic power will be generated by renewable technologies.

In turn, energy-hungry companies will be compelled to maximise efficiency and switch to clean power generation. Renewables will become even more affordable. As a result, the developing world—including China and India—will be able to afford clean power rather than filthy coal.

With a little help from you, right now, the developing giants of Asia might even avoid the full carbon catastrophe which we, in the industrialised world, have created for ourselves.

As these challenges suggest, we are all fated to live in the most interesting of times, for now we are the weather makers, and the future of biodiversity and civilisation hangs on our actions.

I have done my best to provide a user's manual for the climate of planet Earth. Now it's over to you.

acid rain Burning coal with lots of sulphur in it adds sulphuric acid to rain. The resulting acid rain can kill forests, lakes and streams.

aerosols Tiny particles that float in the atmosphere.

albedo Albedo means whiteness. It tells how bright something is and therefore how much sunlight it reflects to space.

Anthropocene A proposed new geological period, defined by human interference with the climate system.

biodiversity All life on earth, including its genetic diversity and ecological interconnectedness.

biofuels Fuels derived from living things (particularly plants), as opposed to fossil fuels.

biomass The total mass of living things in a given time and space—e.g. the biomass of your garden could be calculated (dogs and yourself included).

biosphere The part of our planet that supports life. Generally it is considered to extend to at least 11 kilometres up into the atmosphere, and about the same distance down into the ocean depths.

carbon budget The amount of carbon (or carbon equivalents) that can be emitted into the atmosphere before a threshold is reached.

carbon footprint The amount of carbon that people, industries or countries emit as they go about their business.

carbon sinks Regions or organisms that draw CO_2 from the atmosphere.

carbon tax A tax on CO_2 emissions. It would make industry seek innovative ways to reduce their pollution.

carbon trading The buying and selling of permits allowing people to emit carbon into the atmosphere.

CFCs Manufactured chemicals that destroy the ozone layer. They are also powerful greenhouse gases.

clathrates Methane trapped in ice crystals. They are common on the ocean floor.

climate change Changes to the climate system as a result of global warming.

climate modelling Using computers to project how our climate may change in the future.

climate variability The degree to which climate varies over a given time.

CO_2 emissions Carbon dioxide released into the atmosphere. It may come from industry, plants, the ocean, volcanoes or other sources.

contrails Trails of water vapour generated by jet aircraft. They can turn into clouds.

cryosphere The frozen parts of the earth—the poles, for example.

ecosystem. The interconneted web of life, or a part of it.

El Niño The drought phase of the southern oscillation, which is a cycle bringing drought and flood to large parts of the earth, particularly Australia and South America.

emissions trading A system whereby polluters can trade the right to pollute, allowing those who can most cheaply reduce pollution to do so.

ethanol A kind of alcohol derived from plant matter, and which can be used as a transport or heating fuel.

fossil fuels Typically coal, oil and gas. These fuels are the fossil remains of organisms that lived millions of years ago.

geosequestration The storing of CO_2 in the earth's crust.

geothermal power Power derived from tapping heat in the Earth's crust.

global dimming The cooling of Earth's surface through air pollution or the natural release of certain compounds into the atmosphere.

global warming The warming of Earth's surface through air pollution or the natural release of greenhouse gases into the atmosphere.

green power The generation of electricity without emitting greenhouse gases.

greenhouse gases Gases that trap heat close to Earth's surface. There are around 30 greenhouse gases, of which CO_2 is the most important.

Gulf Stream An ocean current in the North Atlantic that brings heat to Europe.

hybrid-fuel vehicles Vehicles that have both an electric and fuel-driven (typically petrol) engine. The electric engine captures energy (such as that generated when braking) which is normally wasted.

hydro energy Electricity generated from flowing water.

hydrocarbon molecules Molecules composed of hydrogen and carbon. The long chain hydrocarbons (such as petrol and jet fuel) produce a lot of energy from a small volume. Because transport fuels need to be carried in a transport vehicle, such powerful fuels are valuable for transportation.

Jet Stream A current in the atmosphere which is important in influencing the weather in North America.

Kilowatt Enough electricity to run a small home.

La Niña The flood phase of the southern oscillation (see El Niño).

magic gates Times when the world's climate shifts from one stable state to another.

Megawatt 1000 kilowatts. Enough electricity to run 500 large family homes.

methane A molecule made up of 4 hydrogen atoms and one carbon atom. It makes up 90 per cent of natural gas, which is the least polluting of the fossil fuels.

natural gas A gaseous fossil fuel that is around 90 per cent methane.

nuclear energy Energy derived from using radioactivity to boil water.

ozone hole A zone in the stratosphere which is depleted in ozone, and which develops annually over the south and north poles.

photovoltaics A technology which derives electricity directly from sunlight.

power grid The system used to get electricity from the power plant to your home.

renewable energy Energy derived from sources such as wind and the sun, which is effectively limitless.

solar power Power derived directly from harnessing the sun's energy.

solar-thermal Technologies which yield both electricity and heat energy directly from the sun.

sustainability Technologies and lifestyles that give us all a long term future.

Sverdrups A measure of oceanic current flow. A Sverdrup is 1 million cubic metres of water per second per square kilometre.

telekinesis Action at a distance without a visible means of connection.

the carbon cycle A cycle describing how carbon passes through living things, the atmosphere, oceans and Earth's crust.

tidal power Electricity derived from using tidal flows.

wind power Electricity derived from the wind.

FIND OUT MORE

You can always read Tim Flannery's first book about climate change, *The Weather Makers*.

Also check out www.theweathermakers.com for news, updates and online resources. You will also find useful Notes for Teachers and Students.

On the Links and Resources page, you'll find sources that helped inform the views in *The Weather Makers*. There are also links to the Carbon Offset Calculator, the debate in the press on climate change, and a Science Community Forum.

The following websites provide more information about climate change.

YOUR HOME

Origin Energy – information about energy efficiency and sustainability, as well as an energy efficiency calculator and energy saving tips. Follow the links and find out about, for example, Sliver, the new technology of photovoltaics. The Home Energy Project can be ordered online – an excellent resource for use in school and home. www.originenergy.com.au

Sustainable Energy – how to save energy or use renewable energy in your home. The link to schools also provides a good summary of the issues, diagrams and further links. www.sustainable-energy.vic.gov.au

Planet Slayer – this site has a greenhouse calculator. One of the first irreverent environmental websites. www.abc.net.au/science/planetslayer

Greenhousesgases – tips for understanding your energy bill. Tools to help you use less energy, save money and reduce greenhouse gases in our environment. www.greenhousegases.gov.au

Your Home – a comprehensive guide to building, buying or renovating a comfortable, energy efficient home. www.greenhouse.gov.au/yourhome

Energy Rating – explains the energy star rating label program for electrical appliances and includes a database of appliances and their energy ratings. www.energyrating.gov.au

Archicentre – provides architectural services to home buyers, new home builders and renovators. www.archicentre.com.au

YOUR CARS

Smog City – an interactive air pollution simulator that shows how human choices, environmental factors and land use contribute to air pollution. Choose your weather, population level and number of SUVs on the road and see how it affects the smog in this virtual town. www.smogcity.com

Fuel Economy – from the US EPA. See how hybrid cars work, explore alternative fuel vehicles and energy-efficient technologies, ethanol-gasoline vehicles and how fuel cells work. At Find and Compare Cars you can chart how your car affects climate change. www.fueleconomy.gov

YOUR COMMUNITY

Sustainable Living Foundation – a community based not-for-profit organisation committed to promoting sustainable living. www.slf.org.au

TravelSmart – aims to reduce dependency on cars and provide sustainable travel alternatives. www.travelsmart.vic.gov.au

Earthdaynetwork – a great campaign site: links to climate change events around the world, as well as Actions to Combat Climate Change. www.earthday.net

EcoBuy – a program that supports local government and businesses in buying environmentally friendly products. www.mav.asn.au/ecobuy

Climate Friendly – work out how to make your home, school or office 'climate friendly', energy efficient, sustainable. You can buy 'climate neutral' gifts for your friends. www.climatefriendly.com

FIND OUT MORE

STUDENTS AND TEACHERS

National Geographic – links to relevant articles and pictures on climate change. http://magma.nationalgeographic.com
On the National Geographic Adventure magazine site see pictures of Tanzania's Mt Kilimanjero crumbling. There's also a quiz.

World View of Global Warming – photos and links to different world cultures and habitats affected by global warming. Look at the pictures of the Alaskan town of Shishmaref as its shoreline erodes and houses disappear into the sea. www.worldviewofglobalwarming.org

The British Antarctic Survey – press releases and news stories on temperature increases and cryosphere changes. www.antarctica.ac.uk

CH4 – a UK website linking universities, schools, local authorities, NGOs and the public. It includes an online survey. www.ch4.org.uk

Hot Rock Energy – diagrams and explanations about this energy source. Also links to current projects internationally. http://hotrock.anu.edu.au

NASA – great pages for kids, students and adults, information on melting icesheets and hurricanes. www.nasa.gov

Greenhouse Australia – energy saving tips and programmes for school communities, as well as an online photo library and satellite imagery of landscape and vegetation change in Australia. www.greenhouse.gov.au

Climate Action Network – provides up-to-date information on the Kyoto Protocol, geosequestration, new technologies and current politics. www.cana.net.au

EnviroSmart – using the mining industry as a model, students and teachers are guided along a series of school audits in Waste, Water, Land, Air and Energy to become an EnviroSmart school. www.minerals.org.au

I Buy Different – encourages young people to make a difference by the choices they make in what they buy. Links to students' inventions. www.ibuydifferent.org

NOVA Science in the News – The Academy of Science site has links to a range of scientific issues including global warming, climate catastrophes and the carbon currency. www.science.org.au/nova

Planet Ark – includes a picture archive. Find out about their campaigns to help people reduce their damaging impact on the environment. www.planetark.com.au

Re-energy – information and graphics on renewable energies and other clean technologies. You can even build your own working models of renewable energy sources. www.re-energy.ca

Millennium Kids – young people encouraging others to be active about the environment. www.millenniumkids.com.au

Environment Protection Authority (EPA) – provides greenhouse information for students. Follow the links to the Greenhouse Calculator and the Eco Footprint. www.epa.vic.gov.au/greenhouse

Koshland Science Museum: Global Warming Facts and Our Future – includes an online exhibition, interactive graphs showing past and projected climate change, and guides to the greenhouse effect, the carbon cycle and the history of our climate. www.koshland-science-museum.org

BBC Weather Centre: Climate Change – a comprehensive guide to climate change, including audio interviews with key players. www.bbc.co.uk/climate

International Climate Bank and Exchange: Carbon for Kids – includes a useful diagram emphasising the scale of carbon dioxide emissions in the United States. www.icbe.com/carbonforkids

Ollie's World – a child-friendly resource centre with information on issues relating to sustainability principles. www.olliesworld.com

WWF Australia – an independent foundation that promotes environmental education and raises awareness of issues like climate change. www.wwf.org.au **WWF international** – www.panda.org

NOVA Online: Warnings from the Ice – explore how Antarctica's ice has preserved the past – from Chernobyl to the Little Ice Age – and then see how the world's coastlines would recede if some or all of this ice were to melt. www.pbs.org/wgbh/nova/warnings

Globe – schools measure aspects of their environment, and report their results over the Internet. Scientists use GLOBE data in their research and provide feedback to the students. www.globe.org.uk

Climate Change Education – A website hub dedicated to this topic and links to science centres all around the world, a Hot Spots global warming map, links to hands-on science, PBS television shows and to the US Global Change Research Office. www.climatechangeeducation.org

United Nations – Go to CyberSchoolBus then to Briefing Papers. Choose Climate Change for the latest information. www.un.org/english

Commonwealth Scientific and Industrial Research Organisation – click on Energy to find useful information. www.csiro.au

Golden Frog – series of great photos of this extraordinary creature. www.arkive.org/species/GES/amphibians/Mantella_aurantiaca

Antarctic Iceshelves: Larsen B Collapses – satellite imagery and an animation depicting this catastrophe. http://nsidc.org/iceshelves

Coral Bleaching – sponsored by the Australian Academy of Science, this site gives you a good explanation of the phenomenon, focusing on the Great Barrier Reef. www.science.org.au

The Hadley Centre – up-to-date information on weather, climate change and global warming, climate predictions, the carbon cycle. Graphics of charts and other diagrams illustrating trends. www.meto.gov.uk

BUSINESS AND AGRICULTURE

Australasian Emissions Trading Forum – information on emissions trading policy and market developments. www.aetf.net.au

Energy Star – an international standard for energy-efficient office equipment like computers, printers, photocopiers, home electronics like TVs, VCRs, audio products or DVD players. www.energystar.gov.au

Business Sustainability Initiative – Business Victoria's sustainability pages offer information and advice on current Government partnership programmes, grants and assistance. www.business.vic.gov.au/sustainability

The Climate Group – an independent, nonprofit organisation dedicated to advancing business and government leadership in this area. www.theclimategroup.org

Coal 21 – details the advanced technologies being used to reduce and eliminate greenhouse gas emissions from coal. Information on the potential of Zero Emissions Technology. www.coal21.com.au

Cooperative Research Centre for Greenhouse Accounting – provides national leadership in greenhouse accounting research. www.greenhouse.crc.org.au

Greenhouse in Agriculture (GCCA) – this project forms the core of the non-CO_2 (methane, nitrous oxide etc) program for Greenhouse Accounting. www.greenhouse.unimelb.edu.au

GENERAL

Greenpeace – this website will galvanise you into action and inform you about climate change. www.greenpeace.org

FIND OUT MORE

Australian Greenhouse Office (AGO) – provides a range of information on the National Greenhouse Strategy, greenhouse gas emissions and programs to reduce emissions. www.greenhouse.gov.au

Sierra Club – America's oldest grassroots environmental organisation. Recent environmental news stories and commentary. www.sierraclub.org

United Nations Framework Convention on Climate Change (UNFCC) – provides a one-step source of news, data, information and documents on the framework, and related issues such as greenhouse gas emissions and the Kyoto Protocol. http://unfccc.int

The Sun Cube – based on the New Inventors prize-winning solar energy device, the Sunball. Check out how it works, its installation and pictures. www.greenandgoldenergy.com.au

Solar Electric Power Association – everything to do with solar power, especially photovoltaics, and includes a video clip on solar power. www.solarelectricpower.org

Australian Conservation Foundation (ACF) – a leading national, environmental non-profit organisation. www.acfonline.org.au

GREEN POWER

Here is a list of suppliers of Green Power in the UK.

Ecotricity
http://www.ecotricity.co.uk

Green Energy (UK)
http://www.greenenergyuk.com

Good Energy
http://www.good-energy.co.uk

London Energy – Green Tariff
http://www.london-energy.com/showPage.do?name=homeenergy.switchBrad.green.til

Northern Ireland Electricity – Eco Energy
http://www.nie.co.uk/nieenergy/ecoenergy

npower Juice
http://www.npower.com/greenelectricity

PowerGen GreenPlan
http://www.powergen.co.uk/pub/Dom/A/ui/Residential/GreenPlan.aspx?id=60

Scottish and Southern Energy Group – RSPB Energy
http://www.rspbenergy.co.uk

Scottish Power & MANWEB – Green Energy
http://www.warminside.co.uk

Seeboard – Green Tariff
http://www.seeboard-energy.com/showPage.do?name=homeenergy.switchBrand.green.til

SWEB – Green Tariff
http://www.sweb-energy.com/showPage.do?name=homeenergy.switchBrand.green.til

Utilita
http://www.utilita.co.uk

ACKNOWLEDGMENTS

Thanks to Penny Hueston, who shaped *We Are the Weather Makers*, to Alex Szalay for her invaluable input, and to Terry Glavin for drawing my attention to the plight of British Columbia's biodiversity.

ILLUSTRATIONS

Grateful acknowledgment is made to the following for permission to reproduce illustrative material:

BLACK & WHITE

All line art redrawn by Tony Fankhauser: p. 119 from information supplied by the Water Corporation, WA; p. 154 from information supplied by IPCC; p. 189 and p. 204 from information supplied by the Met Office.

p. 2 Star Mountains, New Guinea; p. 6 *Matanim cuscus*; p. 163 *Dingiso*: all copyright Tim Flannery.
p. 84 Tim with giant woolly rat: copyright Gary Steer.
p. 89 *Emperor penguins*: copyright Sharon Chester.
p. 101 *Gobiodon* Species C: copyright Glenn Barrall.
p. 109 Gastric brooding frog: copyright Michael J. Tyler.
p. 129 Hurricane Katrina: copyright NOAA.
p. 167 Lumholtz's tree kangaroo: copyright Karen Coombs.
p. 146 Comparison of climate predictor model and actual weather: copyright The Met Office, UK.
p. 233 Wind farms in Queensland: copyright Stanwell Corporation Ltd.

INDEX

PENGUIN POLITICS

HEAT
HOW WE CAN STOP THE PLANET BURNING

GEORGE MONBIOT

'A dazzling command of science and a relentless faith in people' Naomi Klein

Started to worry about just how hot our world is going to get, and whether you can do anything about it? What with the excuses, the lies, the fudged figures, the PR greenwashing and the downright misinformation on the power of everything from wind turbines to carbon trading, when it comes to saving the world most people don't know what to they're talking about. Luckily, George Monbiot does. Packed with killer facts and inspiring ideas, shot through with passion and underlined by brilliant investigative journalism, with a copy of *Heat* you really can protect the planet.

'I defy you to read this book and not feel motivated to change' *The Times*